Catching Up with Techology

Catching Up with Technology

Published by Quantum Scientific Publishing a division of Sentient Enterprises, Inc. Pittsburgh, PA. Copyright © 2017 by Sentient Enterprises, Inc. All rights reserved.

Permission in writing must be obtained from the publisher before any part of this work may be reproduced or transmitted in any form or by any means, electronic or mechanical, including photocopying and recording, or by any information storage or retrieval system. All trademarks, service marks, registered trademarks, and registered service marks are the property of their respective owners and are used herein for identification purposes only.

table of CONTENTS

Chapter 1 – Introduction to Engineering 2

Chapter 2 – Aeronautics and Aerospace Engineering 5

Chapter 3 – Agricultural Engineering 8

Chapter 4 – Automotive Engineering 11

Chapter 5 – Biomedical Engineering 14

Chapter 6 – Chemical Engineering 17

Chapter 7 – Environmental Engineering 19

Chapter 8 – Industrial Engineering 21

Chapter 9 – Marine Engineering 23

Chapter 10 – Mechanical Engineering 26

Chapter 11 – Nuclear Engineering 28

Chapter 12 – Optical Engineering 31

Chapter 13 – Software Engineering 34

Chapter 14 – Structural Engineering 36

Chapter 15 – Systems Engineering 39

Chapter 16 – Introduction to Biotechnology 44

Chapter 17 – Pharmacogenomics 47

Chapter 18 – Pharmaceutical Products 50

Chapter 19 – Genetic Testing 53

Chapter 20 – Gene Therapy 57

Chapter 21 – Human Genome Project 60

Chapter 22 – Cloning 63

Chapter 23 – Agriculture Improved Crop Yield 66

Chapter 24 – Agriculture and Environmental Stress Reduction 70

Chapter 25 – Crop Nutritional Quality 73

Chapter 26 – Improved Taste, Texture and Appearance of Crops 76

Chapter 27 – Reduced Dependence on Pesticides, Fertilizers 79

Chapter 28 – Production of Novel Substances in Crop Plants 82

Chapter 29 – Biological Engineering 85

Chapter 30 – Biodegradation 88

Chapter 31 – Introduction to Information Technology 90

Chapter 32 – History of Information Technology 93

Chapter 33 – Branches of Information Technology 96

Chapter 34 – Information Security 98

Chapter 35 – Networking 100

Chapter 36 – World Wide Web 102

Chapter 37 – Local Area Network 105

Chapter 38 – Networking Hardware 108

Chapter 39 – Data Management 111

Chapter 40 – Data Storage 114

Chapter 41 – Programming Languages 118

Chapter 42 – Cryptography 121

Chapter 43 – Telematics 123

Chapter 44 – Leaders in Information Technology 125

Chapter 45 – Job Opportunities in Information Technology 129

Chapter 1 – Introduction to Engineering

Chapter Objective:

- Understand the significance of science in engineering

Application of Scientific Principles to Engineering Problems

Engineers design and build solutions to problems, but the solutions to engineering problems cannot be solved by trial and error. For example, civil engineers cannot erect a bridge and wait to see if it fails with the passing of the first traffic. Instead, engineers must understand the scientific principals underlying the stresses placed on various components of the bridge and the materials used in its construction to predict not only whether it will not collapse, but also the weight it will be able to support.

Branches of science underlying the construction of bridges, highways, buildings, dams, and other large structures include mechanics, materials science, fluid mechanics, and mathematics. Failure to understand and calculate all of the forces acting upon a structure can result in disasters such as the spectacular failure of the Tacoma Narrows Bridge, popularly known as "Galloping Gertie." The force of the wind passing over and under the bridge caused the bridge deck to twist, resulting in excessive force on the suspension cables. When the cables failed, the center span of the bridge collapsed into the river below. Amazingly, no one was killed.

Engineering Solutions Drive New Scientific Investigations

Engineering solutions can also lead to new scientific investigations. As computer circuits have become smaller, computer engineers are reaching the limits of classic electromagnetism and must now begin to consider quantum physics.

Physicists and theoretical chemists are developing new materials and new methods of storing information that take advantage of quantum physics while at the same time they must overcome some of the disadvantages. For example, electrons may "leak" or "tunnel" from one side of a resistor to another resulting in an electric current that signals an erroneous change in stored information. Changes in the characteristics of the materials used to store data and the circuits used to carry current in the circuit can reduce or eliminate these errors. Scientists are currently studying the factors affecting the movements of individual electrons to learn how to control individual electron movement.

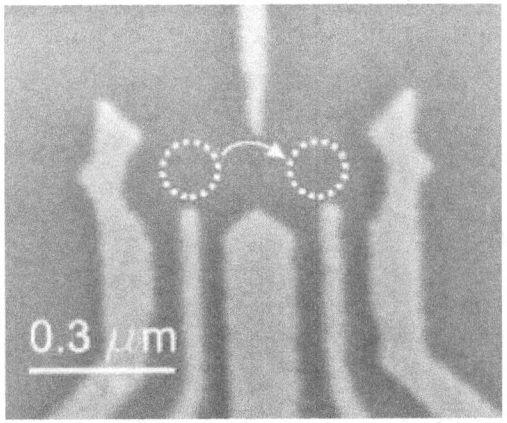

An electron microscope shows the structure of the circuit (Image: Frank Koppens)

Scientific and Engineering Concepts are Expressed Using Mathematics

Mathematics is both a science in and of itself, and the language of all other sciences. One of the more important concepts in electrical engineering and in the physics of electricity is that the current (A) that passes through an object is equal to the voltage (V) applied to the object divided by the electrical resistance (R) of the object. This concept is expressed mathematically by the equation:

$$A = \frac{V}{R}$$

This equation helps us to understand what happens when a short circuit occurs. A short circuit occurs when current flows along a path other than the one intended. Most often, short circuits refer to circuits that are a result of a broken wire or one that has lost its insulation touching an object allowing the current to flow to the ground. In the ground case, the resistance becomes zero. When zero is inserted into the equation above, the result is that no matter how small the voltage applied to the object is, the current becomes, at least theoretically, infinite. This is why short circuits in household appliances and wiring cause so much damage. As the current flows through the wires they heat rapidly and can cause fires or explosions. Fuses and circuit breakers are installed in houses and appliances to prevent short circuits from causing damage.

Summary

Engineers must have a thorough understanding of the principles of mathematics and physics. They often require expertise in chemistry and biology as well. Engineers who do not fully understand these basic principles may design and build objects that either fail to function at all, or result in catastrophes. Engineering solutions to problems often raise as many questions as they answer. Scientists regularly develop new fields of study arising from engineers' questions about the physical properties of materials. Scientists and engineers communicate much of their knowledge using mathematics as a common language. Mathematics provides a logical framework through which key scientific and engineering principles can be understood.

Concept Reinforcement:

1. A client asks you to design a circuit on a new computer chip that will outperform existing chips. Which scientific disciplines will you need to complete your task?

2. A farmer must apply one ton of fertilizer per acre to his farm. His tractor will cover an acre of ground in 20 minutes. There are 8 nozzles on his sprayer. Create a list of variables and a formula to determine the rate of flow required per nozzle.
Calculate the rate of flow required for each nozzle on the fertilizer sprayer to solve the farmer's problem.

Chapter 2 – Aeronautics and Aerospace Engineering

Chapter Objective:

- Understand the significance that science has in aerospace engineering, and analyze its impact on future technologies in this field

Aeronautical Engineering

Aeronautical engineers design and build aircraft used for a variety of purposes. Commercial airliners, fighter jets, helicopters, missiles, UAV's, gliders, and private aircraft owe their specific properties and characteristics to aeronautical engineers. Aeronautical engineers design, construct, and test aircraft engines as well.

Aeronautical engineers use the sciences of fluid dynamics to study airflow over wings and the bodies of aircraft. Materials science and structural dynamics contribute to the strength and stability of aircraft. Aircraft parts must be lightweight, but sturdy. Imagine the impact on the landing gear of a fully loaded 562,000 pound Boeing 747 striking the runway at over 150 miles per hour.

Computer modeling and theoretical mathematics are used to test new designs long before an aircraft is ever manufactured for a test pilot to fly for the first time. The danger to pilots flying an untested aircraft is extreme. Computer Aided Design (CAD), computer models, scale models, and wind-tunnel testing coupled with extensive mathematical analysis of results reduce the likelihood that an experimental aircraft will fail unexpectedly and endanger the test pilot or those on the ground beneath.

Aircrafts are controlled by structures on their surfaces including rudders, ailerons, and elevators. These control structures change the shape of the surface of the aircraft to change its direction in flight. The control structures are operated by computer systems linked to controls operated by the pilot. Avionics is the computer science of aircraft control systems.

As understanding of the science of fluid dynamics increases, due in large part to advances in computer technology, improved wing shapes and control structures on aircraft are being developed. Discoveries in materials science, in particular the new uses for carbon fiber materials, have resulted in stronger, lighter materials for aircraft construction. The combination of improved wings and lighter stronger airframes will increase the maneuverability of aircraft while increasing range and fuel economy.

Aerospace Engineering

Aerospace engineering grew out of aeronautical engineering as flight technology was extended to include space flight. Aerospace engineers design and construct vehicles for extra-atmospheric flight including rockets, satellites, and the Space Shuttle.

Similar to aeronautical engineers, aerospace engineers have to be concerned with all of the forces that act on a vehicle in the atmosphere. But they also have to be aware of the unique needs of extra-atmospheric flight. For example, there is no air in space. Once the spacecraft has left the atmosphere, wings do not generate lift on a spacecraft as they do on an aircraft. It must also carry a supply of oxygen so it can continue to burn fuel once it reaches space.

Aerospace engineers design, construct, and test rocket engines for space travel. Rockets are different from jets. Jets draw in air, compress, and accelerate the air and force it out the rear of the engine to generate thrust. Rockets, on the other hand, burn fuel and force the resulting gases out of the rear of the engine to generate thrust. Rockets require considerably more fuel than jets. Neither jets nor rockets push against the air. Rockets could not function in space where there is no air if this were the case. Instead, they both use Newton's Third Law of conservation of momentum. As mass, either air or combustion products, are accelerated out the rear of the engine, an equal but opposite reaction generates forward thrust.

Spacecrafts are also exposed to extreme heat during atmospheric transit, harsh solar radiation in space, and fast moving physical objects such as meteors. Aerospace engineers must design mechanisms to protect delicate instruments from the effects of heat, radiation, or impact. Many new and unique materials were developed by scientists as a result of the need to protect a spacecrafts' instruments and cargo. Teflon is a common household substance that was developed as a result of space exploration.

Summary

Aeronautical and aerospace engineers use sciences such as mathematics, fluid dynamics, computer science, avionics, materials and science to design, construct, test, and maintain aircraft and spacecraft. Aeronautical engineers have to be concerned with the effects of the atmosphere on aircrafts and understand how to use the air to their advantage in designing aircraft. Aerospace engineers must not only understand the effects of the atmosphere on a spacecraft, they must also understand what happens to it when it leaves the protection of the atmosphere. Improvements in computer technology and materials science will contribute to lighter, faster, more maneuverable, more fuel efficient aircrafts and spacecrafts in the future.

Concept Reinforcement:

1. Which scientific fields do aeronautical engineers need to know to design, construct and test aircraft?

2. Which scientific fields do aerospace engineers need to know to design, construct and test spacecraft?

3. Which scientific advances are contributing to the enhancement of aerospace technology?

Chapter 3 – Agricultural Engineering

Chapter Objective:

- Understand the significance that science has in agricultural engineering, and analyze its impact on future technologies in this field

Agricultural engineers are responsible for designing and constructing agricultural machinery, agricultural structures, irrigation and other water-handling facilities and equipment, and waste management facilities and equipment. Agricultural engineers also design soil and water conservation plans and systems, and conduct environmental studies to develop conservation and remediation plans.

Agricultural engineers must understand mathematics, just as other engineers do, but they must also understand chemistry, biology, soil science, fluid dynamics, structural dynamics, ecology, environmental sciences, and animal and crop production sciences. To understand the complexity of an agricultural engineer's job, let's examine a typical dairy farm.

Dairy farms are among the most complex agricultural enterprises. Dairy farms are integrated agricultural production systems. In other words, dairies are animal production systems, but they usually incorporate crop and pasture systems as well.

Dairy farmers grow corn, soybeans, alfalfa, and other hay crops to feed their cows. Dairy farmers use tractors, plows, harrows, sprayers for fertilizers and pesticides, and a variety of planting and harvesting equipment specific to the crops they grow. Agricultural engineers use mathematics, mechanics, fluid dynamics and crop and soil sciences to design and build the tractors, planters, harvesters, and sprayers that dairy producers need to produce their crops.

Dairy farms have large feed storage facilities such as silos, barns, and cribs to store harvested feed. They use complex mixers to combine the crops and hay into a complete ration for their cows. Modern silos are extremely complex. Because silage is fermented by anaerobic bacteria, silos must be airtight structures that exclude oxygen. Barns and cribs must be designed to keep rodent and insect pests out of harvested feed. Humidity levels must be closely controlled. Too much moisture can result in mold growth and spoilage; too little moisture can dry out feeds and lose nutritional value. Agricultural engineers design structures that allow air to flow through the stored feed to reduce moisture accumulation without drying. They must understand fluid dynamics to control the airflow, but they must also understand plant science to account for the characteristics of the crops being stored and the characteristics of pest species that damage stored feeds.

Cows are milked by milking machines and the milk is transported via a complex plumbing system to bulk milk storage tanks. The storage tanks must chill the milk and keep it stirred or agitated to prevent separation of the cream from the milk. The entire system must be easily accessed for cleaning to prevent the growth of harmful bacteria. Agricultural engineers use fluid dynamics, materials science and mathematics to design and build milking systems. They must also understand the physical characteristics of the cow, especially the udder of the cow to prevent injury to the cow by either excessive pressure on the udder or back-pressure, resulting in milk being forced into the udder rather than being removed.

Cows must be housed and moved easily from bedding areas to feeding areas to the milking parlor. Facilities for treating ill or injured animals are needed. Engineers use structural mechanics to build the facilities, but they must also understand animal behavior to easily move the cows from place to place and provide a comfortable setting in which they can live.

Dairy cows produce a tremendous amount of waste. On average, a dairy cow will produce between 70 and 120 pounds of manure per day. Runoff from manure can create health and environmental hazards, so dairy farmers and other animal producers are required to store and treat manure properly to avoid contamination of the water supply. On a 100 cow dairy, that means farmers must properly store and process 7,000 to 12,000 pounds of manure per day. Engineers use fluid mechanics to design systems that direct manure to storage facilities and prevent runoff. Engineers also use biological sciences to design waste storage facilities that will promote biodegradation of waste products. Agricultural engineers have used environmental science to design artificial wetlands that naturally purify waste products from dairy farms and prevent runoff from contaminating local watersheds.

Future Developments

New developments in the energy industry, including the rapid increase in the price of oil and greenhouse gas production have resulted in a demand for renewable sources of energy that are less polluting. Biofuels, from crops and livestock waste are a new and rapidly developing area of agricultural engineering. Biofuel production requires knowledge of the characteristics of the biological products used to create biofuels, microbiology, and the systems needed to control the biological systems that convert crops and waste into fuel.

Genetically modified crops combine the characteristics of different species into a single species resulting in improved efficiency of production on the farm. However, new characteristics will require agricultural engineers to design new crop handling methods. Environmental concerns about genetically modified crops result in agricultural engineers designing crop production systems that limit exposure of the environment to genetically modified crops. Agricultural engineers must understand genetics and gene transmission among organisms to be successful.

Summary

Agricultural engineers combine their knowledge of mathematics, physics, and chemistry with biological sciences such as microbiology, ecology, environmental science, plant science, animal science, and soil science to design and build complex agricultural equipment and facilities to help farmers produce, store, and transport food. Environmental concerns, and advances in biofuels and genetics are creating new fields of endeavor for agricultural engineers.

Concept Reinforcement:

1. Which sciences do agricultural engineers need to know to design and build crop management equipment?

2. Which sciences do agricultural engineers need to know to design and build animal management systems?

3. What new scientific fields will agricultural engineers have to understand in the future?

Chapter 4 – Automotive Engineering

Chapter Objective:

- Understand the significance that science has in automotive engineering, and analyze its impact on future technologies in this field

Introduction

Automotive engineers are responsible for designing and building motorcycles, cars, trucks, buses, and other motor vehicles. Automotive engineers use mathematics, aerodynamics, computer science, electrical, and safety sciences in designing and building automobiles.

Vehicle Design

Automotive engineers begin by designing a vehicle. They must consider the desired characteristics of the vehicle and the vehicle's purpose. Luxury cars require different suspension systems from heavy trucks, for example. Luxury cars provide a quiet, gentle ride that isolates the passengers from the external environment. Heavy trucks must support tens of thousands of pounds of cargo, but driver comfort is often compromised. A sports car's suspension is designed to provide maximum control at high speed. Each vehicle has different spring, shock absorber, and strut requirements which must be considered by an automotive design engineer. Design engineers use Computer Assisted Design, mechanics, materials science, and mathematics to design the components of each vehicle.

Vehicle Production

Once a vehicle is designed, automotive engineers use Computer Assisted Design to predict the mechanical needs of each system used in a vehicle. Electronic controls operated by the vehicle's computer must also be designed and tested using computer science, mathematics, and electronic engineering science. They then build and test each component of a new vehicle before it is built and tested yet again. Automotive engineers must consider the available materials for constructing a vehicle. Crumple zones on cars, for example, incorporate materials that are designed to collapse at a specific rate to reduce the force of a crash on the vehicle occupants. On the other hand, the occupants are protected by a cage that is structurally very strong and is designed to maintain its shape under the tremendous stress of an impact.

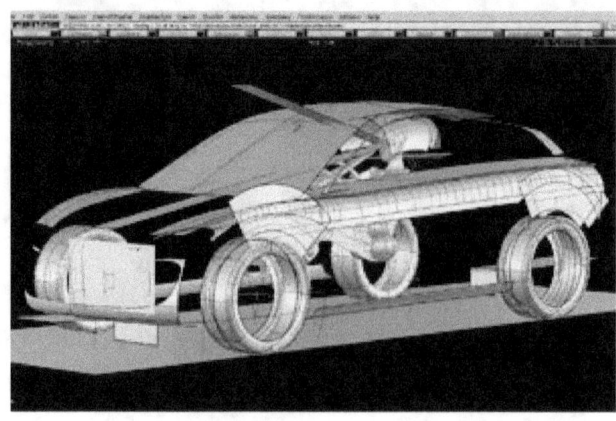

Vehicle Manufacture

The final step in vehicle production is manufacture of the vehicle. Every component of a vehicle must be manufactured, from the smallest electronic controls to the chassis and large body parts. Automotive engineers design the facilities where vehicles are made. Automotive engineers must consider the machines necessary to fabricate a vehicle and its parts, the personnel required to operate the machinery, the speed at which the machinery operates, and the general layout of the plant. They draw on their knowledge of mathematics, computer science, and mechanics to design the auto manufacturing plant. But they must also draw on time and motion studies and labor efficiency studies conducted by human resource and business researchers. Employees must be able to keep up with machinery in the plant and must also be provided safe workspaces.

Future Developments

Sharp increases in oil prices and concerns about greenhouse gases are driving changes in the automotive industry. Automotive engineers are developing and testing new engine designs such as hybrid engines, fuel cell technology, electric vehicle engines, and compressed natural gas engines as alternatives to today's gasoline and diesel engines. New materials such as carbon fiber and lightweight metal alloys are changing the structural design of vehicles and increasing fuel economy by reducing vehicle weight without compromising safety. New computer technology coupled with robotics may soon result in vehicles capable of driving themselves, reducing the likelihood of collisions. Cars of the future will be cleaner, more reliable, safer, and more economical to drive because of automotive engineers' use of new scientific discoveries.

Summary

Automotive engineers design, build, test, and manufacture motor vehicles of all shapes and sizes. From scooters to mining trucks that can carry nearly 500,000 pounds, automotive engineers are involved in every step of production. They utilize their knowledge of mechanics, materials science, computer science, aerodynamics, and human resource management to provide people with reliable, safe transportation. New developments in computer technology, materials science, and chemistry provide the science necessary to develop stronger, safer, cleaner, more economical vehicles in the near future.

Concept Reinforcement:

1. Which sciences are used by automotive engineers responsible for the design of new vehicles?

2. Which sciences are used by automotive engineers responsible for the manufacture of new vehicles?

3. Which sciences are driving changes in automotive engineering?

Chapter 5 – Biomedical Engineering

Chapter Objective:

- Understand the significance that science has in biomedical engineering, and analyze its impact on future technologies in this field

Introduction

Biomedical engineers design and build medical equipment such as MRI machines, medical devices such as catheters, and prostheses, or replacement body parts, for amputees. Biomedical engineers use mathematics, nuclear physics, biomechanics, materials science, biology, pharmacology, biochemistry, electromagnetism, and other sciences to conduct their work. Biomedical engineers perform such a wide variety of functions that it is common for them to specialize.

Medical Imaging

Biomedical engineers are responsible for the design, construction, care, and maintenance of diagnostic equipment in hospitals and large clinics including MRI machines, electrocardiograms, electroencephalograms, X-ray machines and other diagnostic equipment. Biomedical engineers use their knowledge of electronics, electromagnetism, nuclear physics, and medical biology to carry out their duties. Medical imaging devices, for example, allow doctors to see inside a patient's body without surgery. X-ray machines, for example, work by emitting a beam of ionizing radiation through the body. Many of the tissues in the body are incapable of blocking such powerful radiation, and the X-rays reach a sheet of film on the other side of the body to create a photographic image. Some of the tissues in the body, such as bone, are capable of blocking the passage of X-rays, forming a white area on the X-ray film. Radiologists, doctors who specialize in reading diagnostic images, use X-rays and other technologies to see what is happening inside the body.

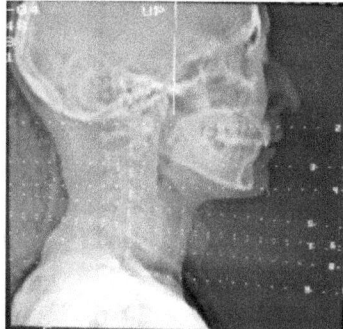

X-ray image of the head and neck

Medical Devices

Medical devices include prostheses, pacemakers, dialysis machines, artificial organs, and instruments used to deliver treatments or provide support to body structures. Biomedical engineers use their knowledge of mathematics, biochemistry, materials science, biomechanics, computer science, electronics, neurology, and other sciences to design and build medical devices. Pacemakers are common devices that are used to control the beat of the heart. Biomedical engineers must understand the function of the heart's electrical impulse system to provide a sufficiently large electrical stimulation to cause the heart to beat in a normal fashion, but not over stimulate it. Batteries must be designed that are small, but powerful and long-lasting to power the pacemaker without the need for frequent surgery to replace the battery. Pacemakers must deliver a pulse of electricity at very specific time intervals for a very long period of time. They must also be adjustable to meet the needs of each individual patient. Small computer chips in the pacemaker must be designed, built, and programmed to ensure proper heartbeat. The materials from which the pacemaker is fabricated must not stimulate an immune reaction from the patient. Clearly, a device as common and simple as a cardiac pacemaker requires a great deal of scientific knowledge from a wide range of scientific disciplines.

Cardiac Pacemaker

Tissue Engineering

The field of tissue engineering by biomedical engineers is still in its infancy. Scientists and engineers are working together to develop replacement organs that can be transplanted into the body from living cells. Scientists seek to understand tissue growth. Engineers apply the new knowledge to growing new organs from cells placed on a collagen or biodegradable scaffold that resembles the final shape of the desired organ. As the cells grow and develop, they break down the scaffold and create their own to result in a completely functional organ ready for transplantation. Engineers must understand cell biology, tissue growth and development, materials science, immunology, and structural dynamics to successfully manufacture artificial organs.

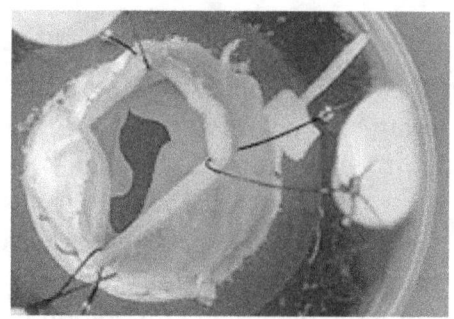

Human urinary bladder grown from amniotic stem cells.

Summary

Biomedical engineers are members of one of the most interdisciplinary of engineering fields. Biomedical engineers work with medical doctors and scientists to design, build, and operate medical equipment, medical devices, and bioengineered tissues. Biomedical engineers must understand mathematics, nuclear physics, biomechanics, materials science, biology, pharmacology, biochemistry, electromagnetism, and other sciences to conduct their work.

Concept Reinforcement:

1. Which sciences are required to design, build, and operate X-ray equipment?

2. Which sciences do biomedical engineers use to design medical devices?

3. Which sciences do biomedical engineers use to engineer new organs from stem cells?

Chapter 6 – Chemical Engineering

Chapter Objective:

- Understand the significance that science has in chemical engineering, and analyze its impact on future technologies in this field

Introduction

Chemical engineers use physics, chemistry, mathematics, and materials science to process raw materials into useful chemicals on a large scale for a variety of industries. Many chemical engineers work in the petrochemical industry, converting crude oil into useful plastics, agricultural chemicals, soaps, cleaners, and other products. Chemical engineers are centrally involved in the design and operation of chemical manufacturing plants to create economical processes of chemical production.

Process Design

Chemical engineers use process design to design chemical factories. They must begin with an understanding to the chemical reactions that must take place to create the desired chemical products. Some products require heat, pressure, or specific pH conditions to form properly. Good process design takes these factors into consideration at the very first stage of design. Chemical engineers usually create diagrams showing the major processes that must occur. The next step involves creating a diagram in greater detail that describes the chemical changes that will occur at each step of the process. Inputs such as heat, pressures, catalysts, and reaction components will be included at this point of the planning. The third step in designing a chemical plant involves drawing plans describing in extensive detail the kind of pipe to be used, the length and diameter of the pipe, and valves needed to control product flow through the pipe. These plan drawings can become extremely complex and confusing and are often broken down into smaller sections for ease of use and understanding. Finally, the specifications of all the major equipment required to operate the plant must be included. The capacity of each component to withstand or create the desired heat, pressure, pH or other condition must be clearly defined as part of the plan, along with energy requirements to operate each component. Chemical engineers write start-up and shut-down manuals for each plant they design. Some reactions are extremely dangerous if the plant is suddenly shut down and can result in explosions and fires that release toxic fumes.

Factors chemical engineers must take into account as they design the chemical plant include facility cost and the cost of maintenance, space availability, safety, waste production and management including potential recycling, environmental regulations, product output including yield and purity, and reliability of the plant. Engineers must thoroughly understand fluid mechanics, thermodynamics, chemical reactions including the potential reaction chemicals in the pipes may have with the pipes themselves, structural mechanics and mathematics to safely design, build, and operate chemical plants. Advances in computer modeling allow chemical engineers to test their designs on a computer before building a plant, reducing costs of construction and potential system failure.

Future Directions

Chemical engineers face many new challenges as large scale chemical production of products, developed from living organisms in bioreactors, are developed. Changes in the energy industry and the availability of oils from plant sources such as soybeans and other oilseed crops for distillation and purification to replace petroleum are requiring new plant designs that operate under different conditions. For example, much of the ink in the daily newspaper is developed from soybeans. Many plastics include animal byproducts from the beef and pork industries. Bacteria have been genetically modified to produce pharmaceutical chemicals that must be extracted and purified in large quantities. Chemical engineers are finding breakthroughs in biology and biochemistry to be a new and growing segment of their profession.

Summary

Chemical engineers are involved in the design and building of large scale chemical processes and the chemical manufacturing plants that produce the desired chemicals. To accomplish their tasks, chemical engineers must have knowledge of mathematics, computer science, basic physics, chemistry, mathematics, structural dynamics, fluid dynamics, and thermodynamics. New areas of growth in the chemical engineering profession will require knowledge of biochemistry, microbiology, plant science and animal science as new, bioengineered products enter the market.

Concept Reinforcement:

1. Which sciences do chemical engineers use to design chemical plants?

2. Which sciences inform chemical engineers of the kinds of pipe they must use in their chemical processing plants?

3. What new sciences are chemical engineers using as new technologies enter the market?

Chapter 7 – Environmental Engineering

Chapter Objective:

- Understand the significance that science has in environmental engineering, and analyze its impact on future technologies in this field

Introduction

Environmental engineers use the sciences of mathematics, chemistry and biology to design and build systems that protect the environment from industrial pollution and safeguard our natural resources through waste treatment and recycling. They must understand the principles of soil science, hydraulics, fluid dynamics, and atmospheric sciences to model the effects of pollutants on the local, regional and global environment. Environmental engineers help industry and government prevent and respond to industrial disasters. They frequently serve as consultants to help industry meet government environmental quality regulations.

Environmental Disaster Recovery

When New Orleans was inundated by flood waters pushed ashore by Hurricane Katrina, more than just water entered New Orleans neighborhoods. Heavy metals such as lead and arsenic, radioactive waste, petrochemicals, agricultural chemicals from Mississippi River sediments, and harmful bacteria from ruptured or flooded sewage lines and facilities are only a few of the hazardous wastes that entered New Orleans with the flood waters.

Due to the epic scale of the disaster, environmental engineers from across the country were recruited to develop and execute the disaster recovery plan. Plans were developed for evacuation of those who lived in the most affected areas. Plans to handle the food, drinking water, and human waste and garbage generated by so many evacuees had to be designed and executed quickly. Water and power distribution plans were put in place, and communications systems had to be installed.

Plans had to be made to remove dead animals, a source of hazardous bacteria. Hazardous chemical spills of uncertain composition had to be contained and removed. Without knowledge of the exact chemical make-up of the spills, it is possible to accidentally combine highly reactive chemicals that can cause a fire or explosion and release of toxic gases. The task of determining where the most likely sites of contamination lay so that clean-up teams could be dispatched was one that required knowledge of fluid dynamics, local topography, and soil properties.

Environmental engineers will be employed for decades developing recovery and remediation plans for New Orleans citizens and the plants and wildlife native to the region.

Future Trends

Environmental engineers are facing new challenges and opportunities in the wake of the growing public concern about global warming and the impact of human activities on the global environment. Finding ways to reduce the emission of carbon dioxide and other greenhouse gases is a rapidly growing field of environmental engineering.

Environmental engineers are also being tasked with developing responses to nuclear, chemical, and biological attacks on populated urban areas. Engineers design plans to reduce the spread of contamination and ultimately remediate the area so that it can once again be used safely by its citizens on a daily basis. Although it may not seem possible, the Japanese cities of Nagasaki and Hiroshima, both of which were destroyed by atomic bombs at the end of World War II, have been completely rebuilt and are thriving, modern, industrialized cities, despite the fact that the radioactive elements left behind by the bomb explosions will last for centuries if not millennia.

Natech is the name of a growing field that has resulted from the combination of <u>nat</u>ural and <u>tech</u>nological combination disasters resulting from events like Hurricane Katrina. The disturbance brought about by a natural disaster is greatly magnified by the technological disaster of hazardous material spills when the natural disaster occurs in a highly populated, industrialized area.

Summary

Environmental engineers use the sciences of mathematics, chemistry, biology, soil science, hydraulics, fluid dynamics, and atmospheric sciences to design and implement plans to protect the environment from industrial and natural disasters. They develop mitigation, recovery, and remediation plans to help industry and government entities meet environmental regulations. Environmental engineers of the future will cope with more complex disasters than ever before because of the growing danger of the combined effect of natural disasters with human activities.

Concept Reinforcement:

1. Which sciences do environmental engineers use to develop plans to protect the environment?

2. In addition to science, what must environmental engineers consider to develop a complete disaster recovery plan?

3. What is a Natech disaster and why is it such a major concern for environmental engineers of the future?

Chapter 8 – Industrial Engineering

Chapter Objective:

- Understand the significance that science has in industrial engineering, and analyze its impact on future technologies in this field

Introduction

Industrial engineering is probably one of the most misleading titles ever given to a professional. Industrial engineers earned the title "industrial" from their early history as experts in designing more efficient manufacturing plants. However, their skills are now used in every business field, from manufacturing, to banking and monetary transactions, to biotechnology labs, to your guest experience at Disney World. Industrial engineers are also called process or systems engineers and design efficiency studies and quality control programs for businesses of all kinds. Industrial engineers use mathematics, statistical analysis, process design and improvement science, simulation modeling, computer sciences, logistics, and human resource management sciences to carry out their duties.

Shortening the Line

Have you ever noticed that stores and banks rarely have all of their lines open at one time? Is it because they don't want to pay another clerk and don't mind inconveniencing you? While that may be what many of us feel while we wait in line, the answer is more complex. For example, if a bank teller can complete 10 customer transactions in an hour, two can complete 20 transactions in an hour and three can complete 30. The bank does not want you to wait for an excessive time in line, but it also does not want tellers standing idle for long period. The bank contracts with an industrial engineer to determine the optimum staffing rate based on the number of customers in the bank per hour and the maximum acceptable waiting period for a customer. If there are seven people in line, how many windows should the bank open? If only one window is open, the last person in line will have to wait 36 minutes before reaching the teller. If two windows are open, the last person in line will only have to wait 18 minutes. If three windows are open, the last customer will wait 12 minutes. The reduction in time spent waiting when there are two tellers versus one teller is very large, but the addition of a third teller reduces the time spent waiting only slightly. Therefore, the bank may choose to operate with only two tellers until the line in the bank grows to 10 people. A tenth person would have to wait 24 minutes if there are only two tellers, but if there are three tellers, that same person would only wait 18 minutes.

Logistics

Industrial engineers also apply their knowledge to logistics, the transportation of goods from place to place. While it may seem to be a simple task to load goods on a truck and transport them to a store, the reality is quite different. Assembly lines in factories are limited by the rate of arrival of raw materials and storage space for those materials, the rate at which the slowest machines are capable of operating, the ability of equipment operators to keep up with the maximum pace of the slowest machines, packaging of the product for shipment, order picking and packing, the rate at which a truck or train can be loaded, the availability of trucks or train cars, the speed with which a truck or train can reach its destination, the rate at which it can be unloaded and ultimately returned to the factory to be reloaded. Each of these processes occurs at different rates and can be slowed or stopped for a variety of reasons. Industrial engineers design the entire system so that as little time as possible is wasted and employees and equipment do not remain idle. Coordinating these efforts over a supply chain that may reach across the globe is a complex task that requires knowledge of human resource management, time and motion studies, complex computer simulations using mathematical models, knowledge of finance and cost analysis, labor and transportation laws, and in many cases international trade practices and regulations.

Summary

Industrial engineers make processes more efficient. They use their knowledge of mathematics, statistical analysis, process design and improvement science, simulation modeling, computer sciences, logistics, and human resource management sciences to design systems with the greatest possible output for the fewest possible inputs. Industrial engineers are involved in nearly every industry in existence. As systems become more complex and more global in their reach, industrial engineers' abilities will be stretched to design efficient systems that keep global trade moving.

Concept Reinforcement:

1. Which sciences do industrial engineers use to accomplish their tasks?

2. In which areas other than heavy industry do industrial engineers work?

3. What factors do industrial engineers have to consider when designing a production system?

Chapter 9 – Marine Engineering

Chapter Objective:

- Understand the significance that science has in marine engineering, and analyze its impact on future technologies in this field

Introduction

Marine engineers are not typical engineers. Most engineers work in an office or occasionally on a construction site. Marine engineers serve as officers on board ships. Marine engineers are directly involved in the operation and maintenance of ships' engines and systems at sea. Sometimes they design and build a ship's system as well, but their primary responsibility is to keep a ship and all its systems running. Not only do marine engineers operate and maintain the ship's engines and propulsion system, they also maintain and operate the ship's water, sewage, electrical, heating and air conditioning, and lighting systems. Marine engineers supervise the ships crew in the operation of all of these systems.

Responsibilities, Knowledge, and Education

Marine engineers have to understand the physics of electricity, high pressure systems, electronics, structural mechanics, chemistry, process engineering, safety science, fire science, and usually first aid and emergency management. Marine engineers also have to make sure the ship is properly balanced when it is loaded so it will not capsize (roll over) or nose under in rough seas. Marine engineers ensure safe refueling. Considering that cargo ships require 50 to 100 or more tons of fuel to travel from port to port, refueling is a major hazardous task. On military vessels, marine engineers may be involved in refueling and operating nuclear reactors, requiring specific knowledge of nuclear physics, nuclear safety and emergency response.

Marine engineers receive their education at naval or marine academies. Military marine engineers train at either the United States Naval Academy at Annapolis, MD, the United States Coast Guard Academy at New London, CT, or the United States Merchant Marine Academy at Kings Point, NY. Marine engineers serving in the military must be able to function well under extreme emergency conditions and prevent a damaged ship from sinking while doing their best to keep the ship functioning as a combat platform.

Merchant fleet marine engineers receive their training at maritime academies located in the Great Lakes, or along the Atlantic, Gulf and Pacific coasts. Upon graduation, marine engineers become third assistant engineers, and can work their way up to Chief Engineer. Each rank requires the engineer to possess the requisite number of hours of experience on board a ship operating the systems for which he will be responsible and pass a licensing exam.

Future Developments

New designs in naval and marine architecture that improve fuel economy, improve handling and maneuverability, and enhance safety are changing the knowledge required of marine engineers. For example, hydroplanes, once only found on pleasure craft, are now included on the PHM1 Pegasus in the U.S. Navy. Stealth technology is being tested on the Sea Shadow, a U.S. Navy test ship. Automated control systems that integrate sensor devices such as radar and sonar with GPS navigation systems are linked to the ship's computer to control the ship's movement without the need for a large crew to operate navigation systems. Cruise ships possess damping devices to reduce the pitch and yaw (up and down and side to side rocking motion) of the ship and increase passenger comfort. Marine engineers help retrofit older vessels with new technologies to extend their useful lifespan.

Summary

Marine engineers serve as officers on board ships and are responsible for the operation and maintenance of ships' engines and systems at sea. Marine engineers supervise the ships crew in the operation of all of these systems. Marine engineers are educated at marine academies operated by military branches of the government or maritime academies operated by universities and private colleges. Marine engineers have knowledge of physics of electricity, high pressure systems, electronics, structural mechanics, chemistry, process engineering, nuclear physics, safety science, fire science, first aid, and emergency management. New technologies to increase speed and maneuverability while increasing fuel efficiency are being developed and tested on military vessels. Those technologies are finding their way onto commercial vessels.

Concept Reinforcement:

1. How are marine engineers different from other engineers?

2. What sciences do marine engineers use in the execution of their duties?

3. What kind of training and experience do marine engineers need to get a job and be promoted?

Chapter 10 – Mechanical Engineering

Chapter Objective:

- Understand the significance that science has in mechanical engineering, and analyze its impact on future technologies in this field

Introduction

Mechanical engineers apply physics and mathematics to solving problems in the design and manufacture of machines, motors, robots, factories, and just about anything else that has moving parts. Mechanical engineers also require knowledge of chemistry, materials science, thermodynamics, statics (the study of non-moving objects under a load), structural dynamics, hydraulics and pneumatics, and fluid dynamics. Mechanical engineers work in the automotive, aerospace, marine, medical, and industrial career fields.

Engine Design

Mechanical engineers begin designing a motor, for example, by determining the power requirements of the system in which the motor will function. An automotive company might decide it needs to build a heavy duty truck that is capable of towing 20,000 pounds, carry a ton of cargo in the truck bed, and accelerate from 0 to 60 mph in 7 seconds. The company will also have fuel economy targets, for example 20 mpg, and size and weight limits for the vehicle itself. Mechanical engineers must then design an engine that will fit under the hood, not cause the truck to exceed its weight limit, and deliver sufficient power to meet load and power requirements. Mechanical engineers must understand the materials that are capable of functioning under the heat and pressure of an internal combustion engine that are light weight but durable. Instead of a steel engine block, for example, they may choose to use an aluminum design that will be lighter in weight.

Determining the power output of the engine requires a complex calculation of the amount of power liberated when fuel is burned in the engine, the amount of space required to burn that fuel, the level of compression that can be accomplished, and the rate of fuel delivery to the cylinders. Engineers use these and other inputs to determine the diameter, or bore, of the cylinders in the engine and the number of cylinders needed. They may also use these calculations to decide whether gasoline or diesel fuel should be used as the power source.

When the specification decisions have been made, mechanical engineers use 3-D Computer Aided Engineering (CAE) or Design (CAD) to diagram the engine and test its components before fabricating any of them. CAE decreases the cost of engine design significantly, because errors can be caught before expensive individual parts are fabricated. Once the engineers are satisfied with the results of their CAE design and testing, they fabricate the engine components and assemble them in a test facility. Engines are tested under a variety of conditions including those that would not normally occur except in an extreme emergency. Only when all of the safety and operational specifications have been fully tested and met will the automotive company begin manufacturing the engine for incorporation into its trucks.

Future Directions

As new materials are developed, mechanical engineers are able to design and build motors and structures that are lighter in weight, stronger, and more durable than ever before. Advances in computer technology, battery design, electronics, and sensory devices such as laser guidance systems and GPS navigation are allowing mechanical engineers to design robots that are capable of performing functions independent of direct human control. Ships, aircraft, and automobiles are becoming increasingly more robotically controlled. These advances require mechanical engineers to update their skills in mathematics, computer science, electronics, and materials science to keep up with new design technologies.

Summary

Mechanical engineers design and manufacture machines, motors, robots, and factories. Mechanical engineers must have a firm grounding in physics, mathematics, chemistry, materials science, thermodynamics, statics, structural dynamics, hydraulics and pneumatics, and fluid dynamics. Changes in materials science, computer science, electronics, and guidance systems have opened new design possibilities to mechanical engineers. But mechanical engineers must also study advances in these and other fields to meet the demands of this ever-changing field.

Concept Reinforcement:

1. Which sciences do mechanical engineers use to design and build machines, motors, robots, and factories?

2. What advances in computer technology have reduced the cost of mechanical engineering design work?

3. How have changes in guidance systems such as GPS and laser guidance systems changed mechanical engineers' jobs?

Chapter 11 – Nuclear Engineering

Chapter Objective:

- Understand the significance that science has in nuclear engineering, and analyze its impact on future technologies in this field

Introduction

Nuclear engineers work with radioactive materials, fusion, subatomic particles, and fissile materials. They have a thorough background in mathematics, including calculus. Nuclear engineers also study nuclear physics, plasma physics, materials science, thermodynamics, fluid dynamics, quantum mechanics, nuclear safety, radiation safety, environmental science, and computer science.

Nuclear engineers design and build nuclear reactors, nuclear weapons, experimental fusion reactors, radioactive waste transport and storage facilities, and radiation detectors. Nuclear engineers are also involved in the design and manufacture of medical imaging devices such as PET scanners, MRI machines, X-ray machines, and radiation therapy devices.

Nuclear Fission

To understand what nuclear engineers do, it is necessary to understand what nuclear fission is. Atoms generally have a specific number of protons and neutrons in their nuclei that give rise to many of their characteristics. However, some atoms of the same chemical element may have differing numbers of neutrons. These atoms of the same element with differing numbers of neutrons are called isotopes. Radioactive atoms are unstable isotopes of non-radioactive atoms. Over time, the unstable neutron is expelled from the nucleus, often releasing energy in the form of heat, or light in the form of X-rays or gamma rays. If a radioactive atom is bombarded with neutrons, the nucleus will split, or undergo fission. The result is the release of large amounts of heat energy and usually several high energy particles including other neutrons. These newly expelled neutrons then collide with more radioactive nuclei, releasing yet more energy and neutrons. The end result is a chain reaction.

In a controlled chain reaction that is not allowed to proceed too rapidly, electrical power can be generated from the heat of the reaction driving steam turbines. An uncontrolled chain reaction results from the detonation of a nuclear bomb. Vast amounts of heat, light and other energy plus large numbers of radioactive materials are released in a very brief times pan causing massive damage to everything in its path. The job of the nuclear engineer is to control the chain reaction in such a way that the desired end result is achieved.

Nuclear explosion

Nuclear reactor core

When enough radioactive material is gathered in a small enough space, the material reaches what is termed critical mass. At critical mass, there are enough neutrons being released by the radioactive material itself that a chain reaction is initiated. In a nuclear weapon, two or more radioactive pellets are rapidly combined, usually by a standard explosive bomb generating a runaway chain reaction and explosion.

In a nuclear reactor, control rods containing a heavy isotope of water are raised or lowered into the radioactive material to either allow more neutrons to interact with the material and speed up the reaction, or to absorb neutrons and slow the reaction. In this way, electrical power can be generated to power nuclear submarines, aircraft carriers, and cities.

Future Developments

One of the most promising nuclear technologies under investigation today is the control of nuclear fusion to generate power. In collaboration with nuclear, quantum and plasma physicists, nuclear engineers are designing and testing facilities capable of containing the vast amounts of energy released by nuclear fusion.

Nuclear fusion differs from nuclear fission in that fission involves splitting radioactive atoms to generate energy and produces radioactive waste. Nuclear fusion involves combining hydrogen nuclei to form helium. Nuclear fusion releases exponentially more energy than fission without creating large amounts of radioactive waste. However, controlling a reaction that is the same as the one that powers the Sun is no small task. It is a major task to simply initiate the fusion reaction. Sustaining and controlling the reaction are even greater problems. However, the reward for achieving the goal of controlling nuclear fusion is an inexpensive form of energy (derived from water) that produces no greenhouse gases and results in little radioactive waste.

Tokamak fusion reactor

Summary

Nuclear engineers work with radioactive materials, fusion, subatomic particles, and fissile materials to design and build nuclear reactors, nuclear weapons, experimental fusion reactors, radioactive waste transport and storage facilities, radiation detectors and medical imaging devices. They have a thorough background in mathematics, nuclear physics, plasma physics, materials science, thermodynamics, fluid dynamics, quantum mechanics, nuclear safety, radiation safety, environmental science, and computer science. The future of nuclear engineering includes the development of a cheap inexhaustible, environmentally friendly electric power supply. However, the challenges to accomplish this goal are great and the science is still in its infancy.

Concept Reinforcement:

1. Which sciences do nuclear engineers use to perform their jobs?

2. What is the difference between nuclear fission and nuclear fusion?

3. What is a chain reaction and how can it be controlled?

Chapter 12 – Optical Engineering

Chapter Objective:

- Understand the significance that science has in optical engineering, and analyze its impact on future technologies in this field

Introduction

Optical engineers design and build devices that use light and its properties to perform useful tasks. To accomplish their goals, optical engineers use physics, especially the science of optics, mathematics, chemistry, materials science, mechanics, and computer science. Products of optical engineers include lenses for microscopes, telescopes and cameras, optical sensors, CD and DVD readers, lasers, and fiber optic communications systems.

Optics in Your Life

As you sit at your computer later reading the online portion of this lesson, stop and think for a moment about the use of light directly under your gaze. The information on your screen was transmitted from a computer over miles of fiber optic cable. Fiber optic cable is designed to use the properties of light to carry information without significant loss of that information. The width of the cable is designed to reinforce the wavelength of the light used to transmit the signal from the computer to your home rather than allow it to disperse, as light from a flashlight in a large tunnel does. Optical engineers are not only able to maintain the beam of light for long distances, but they are able to split the beam into as many as 80 segments within a single fiber using a technique called multiplexing. It is currently possible to transmit 14 Terabits, or 14,000,000,000,000 bits of information over 100 miles on a single fiber. Given the bandwidth required by today's internet users, video on demand cable viewing, and communications, optical engineers are seeking ways to improve even this rapid rate of data transmission.

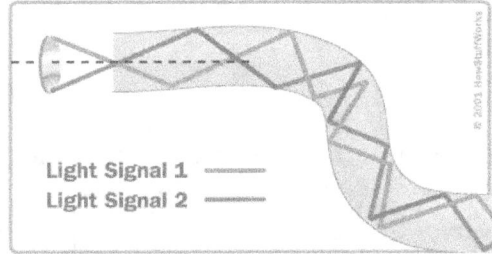

Fiber optic information transmission

You may be listening to a CD while you are working. A CD is really just two layers of plastic with a highly reflective aluminum layer sandwiched in between. However, the plastic is carefully designed by optical engineers to allow transmission of exactly the right wavelengths of light so that only the digital information encoded is read by the optical sensor. A laser in the disc reader reflects from the surface of the aluminum layer to an optical sensor in the CD player. When the laser passes over a pit in the surface of the CD, it takes a fraction of a second longer to reach the sensor than when it reflects from a flat portion of the surface. The difference in whether the laser passes over a pit or the land of the CD digitally encodes the music that was recorded and a program in the CD player converts the digital information into the music you hear. The wavelength of the laser's light is the limiting factor controlling the amount of data that can be stored on a disc. The shorter the wavelength, the more information that can be stored. Most CDs are read using a near infrared laser and can only store up to 700 megabytes (MB). If a blue laser was used, as is the case for Blu-ray DVDs, as much as 50 gigabytes (GB), or 70 times as much information, can be stored on the same sized disc. Blue lasers built by optical engineers have only been available for a short time because scientists have only recently developed the means to generate blue laser light.

The computer chips that perform the processing on your computer are created using a technique called photolithography. Using an x-ray laser, which has a very small wavelength, structures as small as 45 nanometers (nm, 1 billionth of a meter (m)) can be created on a computer chip. Creating the lenses that focus the x-rays through the design mask that creates these tiny structures is the optical engineer's job.

Computer chip structures

Future Directions

The development of new materials with new optical properties has revolutionized optical engineering. Less than a decade ago, fiber optic transmission of data was considered too expensive. Now, there are over 250,000 miles of fiber optic cable carrying information across the US and the world. Developments in materials science have resulted in new coatings for the fiber optic cables, increasing the distance over which information can be transmitted reliably. New lens coatings allow for more specific transmission of desired wavelengths of light resulting in higher quality images from microscopes, telescopes, and cameras. Advances in computer power have resulted in the development of adaptive optics that can reduce or eliminate the twinkling of stars due to atmospheric interference on earth-based telescopes. Advances in computers themselves continue to rely heavily on advances in optical engineering. The future of optical engineering is bright.

Summary

Optical engineers use physics, especially the science of optics, mathematics, chemistry, materials science, mechanics, and computer science to design and build lenses for microscopes, telescopes and cameras, optical sensors, CD and DVD readers, lasers, and fiber optic communications systems. Many common household items are products of optical engineers' efforts. Advances in materials science, optics, and computer science are challenging the field of optical engineering to devise improvements to old technologies and develop new ones.

Concept Reinforcement:

1. Which sciences do optical engineers use to accomplish their goals?

2. Why is the wavelength of light important to consider in the design of new technological devices?

3. What are some common household items that use products designed by optical engineers?

Chapter 13 – Software Engineering

Chapter Objective:

- Understand the significance that science has in software engineering, and analyze its impact on future technologies in this field

Introduction

Software engineers are responsible for designing, evaluating, testing, and building software systems upon which computers operate. They ensure software is written, operated, and maintained in a manner that improves its reliability. Software engineers use mathematics and computer science to accomplish their goals.

Software engineers must keep the goals and desires of the end user in mind. Software designed for an inveterate gamer will be quite different from a business manager's or scientist's data analysis needs. Casual home users may need software that allows them to share family photos and news with friends and family and track the household finances, and little else.

Software Design

Modern computer software may consist of millions of lines of code written by many computer programmers over several years, and include periodic updates to fix problems discovered by users. Software from competing companies must still interact readily with competitor's software on individual computers. Most computer users have experienced the frustration of a software update that creates incompatibilities with other software systems on personal or work computers. Software engineers design systems in such a way as to minimize the likelihood of product failure by standardizing software development procedures. They test new programs for incompatibility with common applications and software platforms. Software engineers are also concerned with evaluating the software for performance issues such as load time, run time, and CPU usage. Fast, efficient programs are preferred to improve consumers' satisfaction with their computers' operating speed.

System Design

Some software engineers design computer systems for government or corporations. They recommend hardware systems to run specific software systems to meet the needs of the company or agency. Banks and credit card companies have high security needs. Software engineers would recommend special firewall software, perhaps on a dedicated server to reduce the likelihood of a customer's accounts being accessed by unauthorized individuals. Universities must balance the need to be open to their students and faculty while at the same time protecting personal information and sensitive research. Universities also face the task of maintaining their computer network security while carrying the burden of teaching new users how to get online and work on the Internet, increasing the risk of computer viruses and other attempts to a hijack a university's computers. Software engineers design separate systems and elaborate security measures to protect a university's sensitive information while allowing open access to information and services.

Future Developments

Software engineering is such a new field that unlike other engineering fields it does not have licensing requirements in most states. Few universities offer degrees in software engineering, and usually offer it as a specialization of computer science degree programs. However, the need for software engineers will drive the development of standards of certification and performance. Software engineering is one of the fastest growing occupations in the US. Increasing computer capabilities resulting from advances in computer engineering and increasingly complex tasks required of computers are driving change in software engineering at a rapid pace. Software engineers will face a plethora of handheld wireless devices that will challenge software engineers to maintain high levels of performance while enhancing a program's efficiency to run on smaller portable devices.

Summary

Software engineers are responsible for designing, evaluating, testing, and building software systems upon which computers operate. They ensure software is written, operated, and maintained in a manner that improves its reliability. Software engineers use mathematics and computer science to accomplish their goals. Software engineers must keep the goals and desires of the end user in mind. Software engineering is one of the fastest growing and fastest changing occupations in the US.

Concept Reinforcement:

1. What sciences do software engineers use to accomplish their tasks?

2. What roles do software engineers play in the computer industry?

3. Why are handheld wireless devices especially challenging to software engineers?

Chapter 14 – Structural Engineering

Chapter Objective:

- Understand the significance that science has in structural engineering, and analyze its impact on future technologies in this field

Introduction

Structural engineers are responsible for the design and building of load-bearing structures. Structural engineers use their knowledge of physics, especially statics, materials science, aerodynamics, hydraulics, and mathematics to create structures that safely and reliably support loads under a wide variety of conditions.

Buildings

Structural engineers design and test the support components of buildings. More experienced structural engineers design and test the integration of these individual supporting components into a complete building. Buildings must meet customer's needs for comfort. Buildings must not sway unacceptably under high wind conditions. Windows must be able to flex without breaking.

Architect's model of Burj Dubai, the world's tallest building currently under construction

Structural engineers design dams which must support the pressure exerted from millions of tons of water. Using their knowledge of shells, dams are often designed with a concave surface so that the force of the water actually increases the structural strength of the dam. Suspension bridges carry thousands of motor vehicles and support their own self- weight on steel cables attached to beams. The calculation of the stresses of weight, vibration from the vehicles, and the aerodynamic effects of wind on the bridge deck must all be modeled on a computer using sophisticated mathematical equations to accurately predict the needs of the bridge in terms of materials used and number of cables and beams required to support the bridge.

Cable stay bridge

Safety

Structural engineers must be especially aware of safety in designing their structures. The catastrophic failure of a building, bridge, dam, aircraft wing, or other structure will almost unfailingly result in massive loss of human life. Most structures are built with redundant safety features to prevent the collapse of a structure without warning. However, unforeseen events can result in the weakening of a building in ways for which the building is unable to compensate. For example, the Oklahoma City bombing of the Alfred P. Murrah Federal Building resulted in a "disproportionate collapse." A disproportionate collapse results when loads and stresses from a damaged portion of a building are transferred to other sections of the building that are not designed to withstand those extra forces. As a result, more of the structure fails than the portion that collapsed as a direct result of the initial damage. When outer structures of the Murrah Federal Building were damaged by the blast, the loads carried by one of the first floor columns were transferred to nearby columns that were not capable of sustaining the new loads resulting in their failure.

Alfred P. Murrah Federal Building

Future Directions

New materials allow structural engineers to create structures that were not considered possible only a few decades ago. Stronger concrete, better steel, improved plastics, and carbon fiber are being used to increase strength while reducing weight. Terrorism, which before the Murrah Federal Building bombing had never been considered in building codes in the U.S,. has changed considerations for the structural design of buildings. Architectural design has become as much art as science, stretching the abilities of structural engineers to meet the artistic inspiration of architects.

Gare do Oriente, Lisbon, Portugal

Summary

Structural engineers use their knowledge of physics, especially statics, materials science, aerodynamics, hydraulics, and mathematics to design and build load-bearing structures. Stronger concrete, better steel, improved plastics, and carbon fiber allow structural engineers to create structures with increased strength and reduced weight. Terrorism has contributed to changes in the structural design of buildings. Architectural design has become as much art as science, stretching the abilities of structural engineers to meet the artistic inspiration of architects.

Concept Reinforcement:

1. Which sciences do structural engineers use to accomplish their tasks?

2. What are some of the factors structural engineers must consider when designing a structure?

3. What scientific disciplines are driving changes in structural engineering?

Chapter 15 – Systems Engineering

Chapter Objective:

- Understand the significance that science has in systems engineering, and analyze its impact on future technologies in this field

Introduction

Systems engineers are interdisciplinary problem solvers. Systems are broadly defined as a group of interacting components that produce effects greater than the sum of the individual components. Systems can be physical systems such as a conveyer system that moves bottles through a bottling plant, computer programs, or management systems that control the flow of product around the globe. Systems engineers design durable, efficient systems that meet the needs of their clients.

Method

Systems engineers use logic, mathematics, and computer science modeling techniques to create solutions for their clients. The International Council on Systems Engineering (INCOSE) developed the process known as SIMILAR which developed solutions for a wide range of businesses:

1. State the problem
2. Investigate alternatives
3. Model the system
4. Integrate
5. Launch the system
6. Assess performance
7. Re-evaluate

While re-evaluate is a specific step of the process at the end of system design, it is a continuous function of the entire process. For example, after modeling the system, systems engineers may find that the model predicts the system will not work. In such a case, they must either develop a new model based on the alternative chosen in the previous step, or they must choose a new alternative and design a new model based on the new alternative.

When designing a system, systems engineers must consider the entire life cycle of the system from initiation, to scale-up, through maintenance and upgrades, to the decommission of the system for replacement with a new one. If maintenance, upgrades, and decommission are not considered in the initial design of the system, the overall cost may become unsustainable and the useful lifespan of the system may be shortened. For example, decommissioning a nuclear power plant is an exceptionally expensive undertaking, and the design of the plant and its operation must take into account this cost, or the company may be driven into bankruptcy during decommission.

Quantitative risk management and decision-based design are two of the mathematically based management sciences systems engineers use to design their systems. Systems engineers produce a decision tree that includes points at which decisions must be made, the outcomes of the decisions, and the statistical probability of achieving the outcome. By including the statistical probability of the decision, a model of the cost and income generated by each decision can be created, making profit optimization decisions easier to visualize.

Probability decision tree

- Outcome 1 — 20% likely
  - Risk B — $30,000 implication
    - Outcome 1.1 — 25% likely
    - Outcome 1.2 — 70% likely
    - Outcome 1.3 — 5% likely
- Outcome 2 — 80% likely

Risk A — $10,000 implication

By multiplying the risk associated with each outcome by the dollar amount, the manager can quickly determine which outcome is the most desirable. If Risk A does not occur, the cost to the company would be $8,000. However, if Risk A does occur, there is a possibility that Risk B will occur. Risk B has three possible outcomes. The formula a systems engineer would devise is:

(Risk A cost*Risk A probability)+(Risk B cost*Risk B probability)= Total cost

The cost of both Risk A and Risk B can be seen in the table below.

Risk A	Risk B	Risk A cost	Risk B cost	Outcome Probability P(A)*P(B)	Total
80%	0%	$10,000	0	80%	$8,000
20%	25%	$10,000	$30,000	5%	$4,500
20%	70%	$10,000	$30,000	14%	$23,000
20%	5%	$10,000	$30,000	1%	$3,500

Managers can easily see that outcome 3 is the outcome of choice as it has the least cost. However, it is also the least likely with only a 1% chance of occurrence. Risk A alone is much more likely (80%), so managers should be prepared for the potential cost of $8,000 to occur.

In more complex systems, such as the development of the US Army Future Combat System, which was completed by systems engineers and includes personnel, equipment, transportation, fire support, and aircraft including UAVs, the decision tree can be exceedingly complex. However, the final outcome of the process is readily available in terms of risk and cost, regardless of the level of complexity.

Summary

Systems engineers are interdisciplinary problem solvers who design durable, efficient systems that meet the needs of their clients. Systems engineers use logic, mathematics, and computer science modeling techniques, formalized as the SIMILAR process, to create solutions for their clients. Systems engineers use the process to develop solutions for a wide range of businesses. Advances in computer science and technology, as well as business risk management science and decision making science make the job of the systems engineer more complex, but at the same time, easier by providing power and structure to the design process.

Concept Reinforcement:

1. Which sciences do systems engineers use to design systems for government agencies and corporations?

2. What parts of a process must systems engineers remember to include in the design of a system?

3. Draw a decision tree and create a table of outcomes for the following set of Risks and Costs.

Risk	Probability	Cost
A	70% yes / 30% no	$300
B	25% yes / 75% no	$1,000
C	10% yes / 90% no	$500

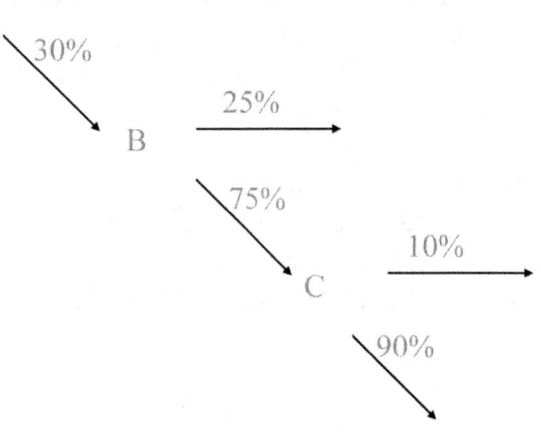

Risk A	Risk B	Risk C	Risk A cost	Risk B cost	Risk C cost	Outcome Probability P(A)*P(B)*P(C)	Total
70%	0%	0%	$300	0	0	70%	$210
30%	25%	0%	$300	$1000	0	7.5%	$340
30%	75%	10%	$300	$1000	$500	2.25%	$890
30%	75%	90%	$300	$1000	$500	20.25%	$1290

Chapter 16 – Introduction to Biotechnology

Chapter Objective:

- Understand the significance of science in biotechnology

Introduction

Biotechnology, according to the United Nations Convention on Biodiversity, is "any technological application that uses biological systems, living organisms, or derivatives thereof, to make or modify products or processes for specific use." This definition is very broad and can include centuries old practices such as wine and beer making, both of which use bacterial fermentation to produce alcohol from plant products. It also includes genetic engineering of organisms in which DNA is inserted into their cells to produce a specific gene product such as insulin. The US Department of Agriculture adds to the definition: biotechnology "…includes recombinant DNA technology, transgenic crops and animals, genetically modified foods, biopharmaceuticals, bioremediation, and more."

Biotechnology is most commonly associated with the agricultural and medical fields, but it can also be applied to engineering fields such as nanotechnology and environmental engineering. Molecular "machines" have been fashioned from proteins to do work at the microscopic level. Environmental engineers may use genetically altered organisms to clean up hazardous wastes that cannot be otherwise removed. Computer engineers are investigating the possibility of using biomolecules for next generation computer hardware.

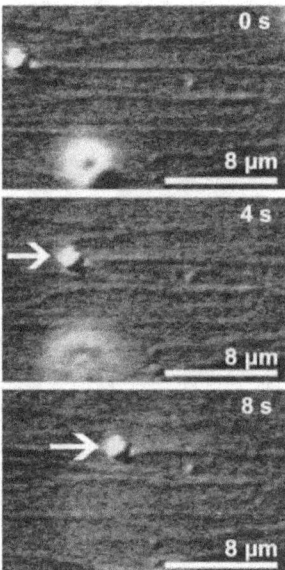

Biomolecular nanomotors move micron-sized balls

Science in Biotechnology

Biotechnology is one of the most rapidly advancing technological industries in the world. The rapid advances in biology, biochemistry, and computer sciences drive changes in biotechnology at a phenomenal pace. A thorough understanding of organisms, their body systems and functions, and how changes in DNA result in changes in the characteristics of the organism as well as how to control those changes are required of any biotechnologist. Microbiologists are especially involved in the biotechnology revolution, since many of the current crop of biotechnology-based products are either produced by, or engineered from bacteria.

The sheer volume of genetic information encoded on a strand of DNA requires knowledge of computer science, statistics, and informatics to store the information in a computer database, retrieve the useful DNA sequences, or comb through all of the data to identify new genes.

Engineering, chemistry, and physics contribute heavily to biotechnology discoveries as well. Understanding how to create chemical linkages without damaging small molecules is critical to designing biomolecules as products. Knowledge of physics and quantum effects is important in understanding and predicting the behavior of molecular machines. Mathematics is especially important. Computation of the forces that act on nanodevices is needed to prevent them from failing.

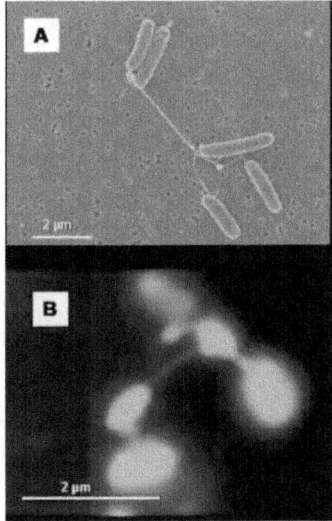

Bacteria produce electrically conductive nanowires, which could be used in next-generation computer hardware.

Genetics and the development of cloning techniques have led to the birth of animals possessing genes from other species. For example Nexia, a biotechnology company, has created goats that possess the gene for spider silk. Their goal is to collect spider silk protein in large quantities from the goats' milk. Spider silk is many times stronger than steel and carbon fibers, yet is lightweight. By producing spider silk in quantity, lighter, stronger products from aircraft to bullet-proof vests could be manufactured.

Genetic modification of crops, involving crop scientists, geneticists, soil scientists, and microbiologists has resulted in increased crop yields, reduced stored crop losses, and helped feed millions. A further benefit is a reduction in the use of pesticides or a conversion to less hazardous pesticides to grow food and oil crops. Over 80% of the soybeans and cotton grown in the US are genetically modified. Over half of the corn grown in the US is genetically modified. Crop scientists anticipate continued growth of genetically modified plantings in the US and the rest of the world.

In medicine, the use of biotechnology has grown dramatically. Genetically modified bacteria produce insulin and other important biomolecules to treat disease. Antisense DNA technology is being developed by biochemists and pharmaceutical chemists to prevent damaged, mutated, or other harmful genes from functioning to treat genetically inherited diseases. New vaccines using "naked" DNA are being developed to prevent disease. Microarray gene chips are being developed to detect and diagnose diseases. The so-called "lab on a chip" that can diagnose disease using a drop of the patient's blood at the patient's bedside is becoming available.

Biotechnology is also being used in antiterrorism devices. Biochemicals bonded to an electrically conductive surface change shape when they bind to chemicals specific to explosives or to harmful bacteria and viruses. The change of the molecules' shape changes the flow of current in the base and sets off an alarm. Chemistry, biochemistry, and electronics combine to make these powerful detectors a potential product of the future.

Summary

Biotechnology is everywhere in our lives. Biotechnologists use the characteristics of living organisms to design and manufacture products in medicine, agriculture, computer engineering, electronics, and nanotechnology devices. Biotechnologists must have knowledge of biology, microbiology, chemistry, biochemistry, genetics, physics, mathematics, computer science, and a host of other sciences specific to their industry to complete their tasks. The horizon for biotechnology is as broad and diverse as life itself.

Concept Reinforcement:

1. Which sciences are important for biotechnologists to know?

2. How are changes in science changing the field of biotechnology?

3. What are some common biotech products that impact your life?

Chapter 17 – Pharmacogenomics

Chapter Objective:

- Understand the significance that science has in pharmacogenomics, and analyze its impact on future technologies in this field

Introduction

Pharmacogenomics is the study of the interaction of the entire human genome and how it controls the body's reaction to drugs and other medical treatments. Pharmacogenomicists use pharmacology, biochemistry, organic chemistry, genetics, physiology, and computer science to predict how drugs will act and prevent adverse side effects in individuals rather than in entire populations as is now the case in pharmaceutical drug development. Side effects are responsible for over 100,000 deaths and well over 2 million serious cases per year in the US alone. Many more patients fail to respond to treatment at all. Pharmacogenomics attempts to "personalize" drug selection based on an individual's genes to ensure the best possible response to treatment without side effects.

Genetic Variation

Take a moment to look at the people around you. Every one of them has a different physical appearance, even though all are clearly members of the same species. The variation in appearance, as well as the difference in their response to the environment, heat, sunlight, and food, is due in large measure to tiny changes in the genetic sequence of their DNA. The genes in your cells are encoded by only four DNA nucleotides. The nucleotides are "read" by the cell's machinery in groups of three, with each group of three coding for a specific protein building block, or amino acid. When one of those nucleotides is changed, the amino acid is also changed. This leads to the differences you can see.

Some of these changes in the genetic code alter how a person reacts to drug treatment. To be effective, most drugs must bind to another molecule and either prevent it from functioning, or improve its function. But if the target protein has a different shape, or if the amino acids at the binding site are replaced by others, the drug may not bind well. On the other hand, it is possible that the drug will bind too well. In either case, an adverse side effect may occur. In some cases, the drug may not bind at all and the patient will not respond to treatment. Pharmacogenomics attempts to predict how well a drug will function in an individual based on his or her DNA.

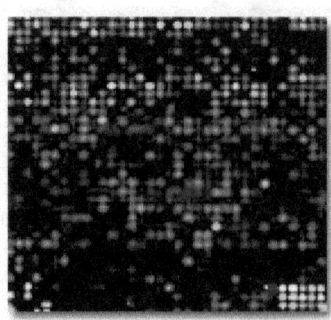

Molecular profile of cancerous tissue using a gene expression microarray. The expression microarray informs doctors which genes are "on" in the cancer cells.

In order to predict whether a drug will be effective for a patient, pharmacogenomicists must understand organic chemistry and biochemistry. Using their biochemical knowledge, they must predict the amino acid composition of the drug target and its three dimensional shape. Using their organic chemistry knowledge, they must predict how the drug will bind to the target and the effect of drug binding on the target.

Pharmacogenomicists must understand computer science and bioinformatics to handle the massive databases that are created by the vast amount of information encoded in the human genome. They scan the billions of DNA nucleotide sequences in the human genome to match the patient's DNA to known gene sequences and determine variations in the patient's DNA that may affect drug function. They must then use computer systems to match drugs to targets for the best possible patient outcome.

Future Directions

The ultimate goal of pharmacogenomics is to design drugs for each individual that will directly target the disease and have no effect on normal cells. The field is still in its infancy, with several important hurdles to overcome. The biggest hurdle is the difficulty of finding differences in the gene sequence that affect a drug's function. There are literally millions of potential genetic differences that could have no effect on a drug's ability to bind to its target properly. For many diseases, there are only one or two drugs available. The treatment options are so limited as to not be worth the time and expense of sequencing the patient's genes. Pharmaceutical manufacturers are not willing to create numerous individualized drugs because of the hundred million dollar cost of bringing a single drug to market and the relatively low profit on most drugs. Creating many specialized drugs is not now economically feasible. Finally, many doctors do not have the training in genetics they would need to use pharmacogenomics effectively in diagnosing and treating their patients. Until pharmacogenomics becomes available for a wider range of diseases and becomes easier to use, its promise of individualized medicine will remain unfulfilled.

Summary

Pharmacogenomics is the study of the interaction of the entire human genome and how it controls the body's reaction to drugs and other medical treatments. Pharmacogenomicists use pharmacology, biochemistry, organic chemistry, genetics, physiology, and computer science to predict how drugs will act and prevent adverse side effects in individuals rather than in entire populations. The ultimate goal of pharmacogenomics is to design drugs for each individual that will directly target the disease and have no effect on normal cells. The field is still in its infancy, with several important hurdles to overcome: 1) the difficulty of finding differences in the gene sequence that affect a drug's function, 2) limited treatment options, 3) economic feasibility, and 4) many doctors do not possess enough genetics training.

Concept Reinforcement:

1. Which sciences are used in pharmacogenomics?

2. What happens when a DNA nucleotide in a gene sequence is changed?

3. What are the goals of pharmacogenomics?

4. What are the barriers to full implementation of pharmacogenomics in medical practice?

Chapter 18 – Pharmaceutical Products

Chapter Objective:

- Understand the significance that science has in pharmacology, and analyze its impact on future technologies in this field

Introduction

Pharmaceutical products are products that have medicinal qualities. They are sought, screened, tested, designed and manufactured by pharmacologists. Pharmacologists study how drugs interact with the body to change its function. They use organic chemistry and biochemistry to understand drug composition. Organic chemistry, biochemistry, physiology, and toxicology are sciences used to study drug interactions. Pathology, microbiology, virology, mycology, genetics, and physiology are used to determine a drug's medical applications and ability to kill disease-causing organisms. Computer science, and mathematics are important in understanding the pharmacokinetics, or useful lifespan of a drug in the body and how the body metabolizes the drug.

Divisions of Pharmacology

Pharmacology is such a broad science that it is subdivided into therapeutic areas. Each area requires specific knowledge in addition to the knowledge required of all pharmacologists.

Pharmacokinetics

Pharmacokinetics is the study of the body's effects on a drug. From the moment a drug enters the body, the body begins to break it down. The body also transports the drug from its initial location to the rest of the body. For example, when a patient swallows a pill, the stomach's acids begin to break down the active ingredient of the drug. Many pills and capsules are coated to prevent early damage by stomach acid. At some point in the digestive tract, however, the pill must be broken down or the medicine cannot be absorbed into the body. Even as the body is absorbing the drug, enzymes in the liver and kidneys begin the process of chemically breaking down the drug for excretion in the urine or feces. Pharmacologists must determine the rate of absorption, the rate of excretion, and the dosage required to have a therapeutic effect to determine the amount of the drug that must be administered to achieve a therapeutic dosage in the body within a specified time period. Pharmacokinetics also studies the chemicals produced by the body at each stage of a drug's metabolism to ensure that none of the compounds are toxic.

Neuropharmacology

Neuropharmacology is the study of drug effects on the function of the nervous system including the brain, and the behavioral effects of those drugs. Neuropharmacologists work with chemicals that either prevent neurons from transmitting a signal, or enhance the transmission or duration of the signal. Neuropharmacologists must understand neuropsychology, the study of the brain's effect on behavior. Neuropharmacological drugs often have behavioral effects. Some behavioral effects are desirable, as in the case of antidepressive drugs. Some effects are undesirable, as in the case of hallucinations resulting from some drugs that are used to treat Parkinson's disease.

Pharmacogenetics and Pharmacogenomics

Pharmacogenetics and pharmacogenomics are sciences that link pharmacology with genetics. In the case of pharmacogenetics, scientists study the effects of single genes on drug effects. Pharmacogenomics, on the other hand, studies the effect of the entire genome on a drug's effect. Both are aimed at increasing the specificity and efficacy of drugs while reducing undesirable side effects. Knowledge of genetics, protein synthesis, mathematics, computer science and bioinformatics are all important to linking drug effects and genes in cells.

Pharmacoepidemiology

Drugs are thoroughly tested by manufacturers and scrutinized by the US Food and Drug Administration (FDA) before they enter the market. However, once a drug enters the market data is collected to study the effects of the drug in a large population. Scientists who collect and study this data are pharmacoepidemiologists. They are responsible for reporting problems that went undiscovered during the initial phases of the drug approval process. Fen-phen, a combination of fenfluramine and phentermine, was pulled from the market when pharmacoepidemiologists discovered it was responsible for damaging heart valves in patients. On the other hand, old drugs are sometimes discovered to have other beneficial effects. For example, Viagra was originally used to treat high blood pressure.

Toxicology

Toxicology is the study of poisons. Poisons have specific effects on the body that cause cellular machinery to function improperly, or not at all. However, many useful medicines have been developed from toxins and poisons. Botox, for example, is a one of the most potent poisons in the world, having neurotoxic effects. However, it is used in the treatment of painful muscle spasms, and of course, is used in cosmetic surgery. Toxicologists study the effects of toxins and poisons on the body and develop ways to neutralize them.

Future Directions

The rapid advances in our understanding of the effects of genes on drugs, and how drugs can be used to manipulate gene function is one of the factors driving change in the pharmaceutical industry. Knowledge of the genes of disease-causing organisms has resulted in the development of new drugs that are more effective and that results in less drug resistance. Intelligent drug design, in which a drug is specifically designed to bind to a chemical in the body to either halt or enhance its function to treat disease, is being made possible by advances in computer science. The combination of these two fields of study is leading to the ultimate goal of individualized medicine in which a drug is specifically designed for an individual based on his or her genetic code to be maximally effective with no side effects.

Summary

Pharmacologists seek, screen, test, design and manufacture drugs. Pharmacologists use organic chemistry, biochemistry, physiology, toxicology, pathology, microbiology, virology, mycology, genetics, computer science, and mathematics to study how drugs interact with the body to change its function. The rapid growth in our understanding of computer science, genetics, and how gene products may interact with drugs is driving rapid change in pharmacology and the development of pharmaceutical products.

Concept Reinforcement:

1. Which sciences are used by pharmacologists in developing pharmaceutical products?

2. Toxicologists contribute to pharmacology in which two ways?

3. Which sciences are driving changes in the pharmaceutical industry and how are they doing so?

Chapter 19 – Genetic Testing

Chapter Objective:

- Understand the significance that science has in genetic testing, and analyze its impact on future technologies in this field

Introduction

Genetic testing is used to diagnose genetic diseases and to determine ancestry. Geneticists can diagnose genetic disease by observing the chromosomes, called gross chromosomal observation, for correct shape and number. They can also diagnose diseases or susceptibility to disease at the level of the individual gene. Each person receives half of his or her genes from the father, and the other half from the mother. Geneticists can use this knowledge to identify parents of children, unidentified bodies, victims of genocide, and migration of prehistoric human populations. Geneticists must have a thorough understanding of genetics, biochemistry, bioinformatics, computer science, and mathematics including statistics.

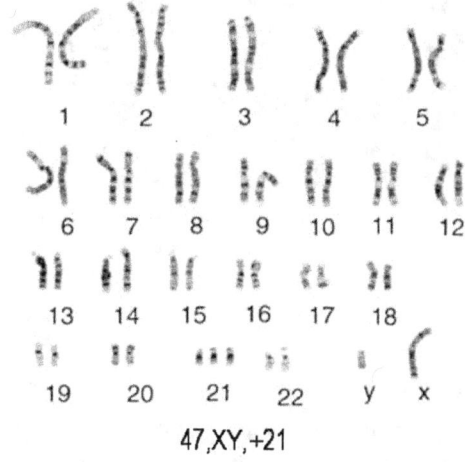

Chromosome squash showing trisomy 21 (Down's Syndrome) in a male.

Gross chromosomal observation

Geneticists have long used gross chromosomal observation as a means of diagnosing disease. For example, geneticists obtain cells of a fetus before it is born using a technique called amniocentesis. Amniotic fluid is extracted from the placenta using a long needle. Some of the fetus' cells will be suspended in the fluid. Geneticists extract the DNA from an actively dividing cell and count the number of chromosomes and examine their shape, or morphology. Down's syndrome, which can result in severe mental retardation and characteristic physical abnormalities, is also known as trisomy 21, or "three body" 21. Instead of the normal pair of chromosome 21, there are three.

Genetic Screening

The ability of geneticists and biochemists to examine individual genes allows individuals to undergo genetic screening. Advances in computer science and optical engineering have helped move the field forward. People with a family history of a genetic disease may choose to be screened to determine the likelihood that they will contract the disease. BRCA 1 is a protein responsible for correcting mutations and damage to DNA. Women who possess a mutated form of the *BRCA 1* gene are more susceptible to breast cancer than other women because mutations cannot be repaired by the mutant form. In some cases the risk of cancer is 3 to 7 times greater than in women who do not have the mutation. This increased risk has caused some women to decide to have preemptive radical mastectomies in which both breasts are surgically removed.

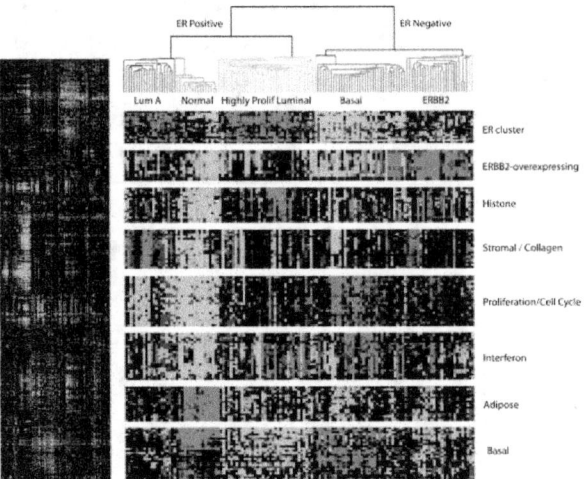

Microarray screening of breast cancer tissue demonstrates five or six molecular subtypes.

Genetic Counseling

Because of the potentially devastating news that one has inherited a disease for which there is no cure, the field of genetic counseling has grown up around genetic testing. No one should be given the results of a genetic screen without a qualified counselor on hand to assist him in understanding what the test actually means. There is no guarantee that the patient will ever contract the disease, but many patients will become depressed, angry, or feel guilty because of their test results. Genetic counseling by a qualified genetic counselor trained in both genetics and psychology can relieve some of the worry associated with the diagnosis of a potential genetic disease.

Forensic Genetic Testing

Genetic testing can be used to identify individuals and their relation to one another. Because offspring receive one copy of their DNA from the father and one copy from the mother, it is relatively easy to establish paternity using genetic testing. DNA samples are collected from the mother, child, and suspected father or fathers. The DNA is processed and separated using gel electrophoresis to determine which DNA samples match.

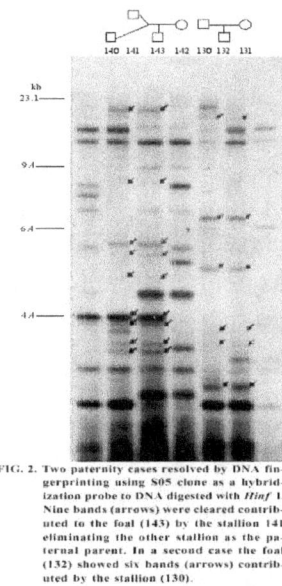

FIG. 2. Two paternity cases resolved by DNA fingerprinting using S05 clone as a hybridization probe to DNA digested with *Hinf* I. Nine bands (arrows) were cleared contributed to the foal (143) by the stallion 141 eliminating the other stallion as the paternal parent. In a second case the foal (132) showed six bands (arrows) contributed by the stallion (130).

Stallion 141 is the father of foal 143, the other stallion does not match foal 143's DNA. Stallion 130 is the father of foal 132.

Genetic testing can also be used to identify victims of crime or disaster. When the World Trade Center in New York City collapsed after the terrorist attacks of September 11, 2001, many of the building's occupants were missing. Families of missing people provided hairbrushes, toothbrushes, and other personal items so that the missing person's DNA could be extracted and examined. When human remains were discovered, DNA was extracted from them and matched to the database of DNA samples from missing persons. Of the 2,749 people who died, 86% were identified using forensic genetic analysis.

Future Directions

The future of genetic testing is plagued by ethical issues. Direct to consumer genetic tests allow people to conduct genetic tests in the privacy of their own homes, but their reliability is not good. As a result, people may make bad decisions based on poor quality results. Pharmacologists and geneticists are working to develop better home tests and government regulatory agencies may take a more active role as well. Privacy of genetic information must be guarded, and computer scientists are working on secure data systems to protect genetic information. Because of the potential for genetic discrimination in health insurance and employment decisions, many people do not get screened.

Summary

Genetic testing is used to diagnose genetic diseases and to determine ancestry. Geneticists can diagnose genetic disease by gross chromosomal observation or at the level of the individual gene. Geneticists can identify parents of children, unidentified bodies, victims of genocide, and migration of prehistoric human populations. Geneticists must have a thorough understanding of genetics, biochemistry, bioinformatics, computer science, and mathematics including statistics. Genetic counselors trained in both genetics and psychology help patients cope with bad results of genetic tests.

Concept Reinforcement:

1. Which sciences do geneticists use to conduct genetic testing?

2. Why would a geneticist need to understand psychology?

3. How have computer sciences changed the field of genetic testing?

Chapter 20 – Gene Therapy

Chapter Objective:

- Understand the significance that science has in gene therapy, and analyze its impact on future technologies in this field

Introduction

Gene therapy involves the insertion of a healthy copy of a gene into the cells of a patient suffering from a genetic disease because the patient possesses a faulty copy of the gene. Gene therapists have a medical degree, and also possess in depth knowledge of genetics, biochemistry, organic chemistry, mathematics and statistics, and physiology. Gene therapy is still in its infancy and faces a number of problems. But advances in virology, biochemistry, immunology, and other scientific fields are addressing many of these shortcomings.

Using Viruses to Cure Disease

Viruses are disease-causing organisms that hijack healthy cells' functions by inserting their own DNA into the healthy cell. The viral DNA directs the cell to start producing viral proteins and viral DNA, making new viruses until the cell ruptures, or lyses. The new viruses then go on to infect new cells. Gene therapists take advantage of the virus' ability to insert DNA into cells. But instead of viral DNA, gene therapists replace the virus' DNA with the gene needed by the patient. When the virus infects the patient's cells, it inserts a normal, or *wild type*, copy of the gene the patient needs and the cell begins to function normally. Unfortunately, viruses have several shortcomings as gene carriers, or *vectors*. There is no way to target the viruses to the cells that need the gene therapy. For example, if the patient has cystic fibrosis, a lung disease, many of the viruses will target skin, liver, heart, or other cells and have no effect on the disease. Viruses are small and human genes are large. It is difficult, if not impossible to fit human genes inside a virus in many cases. Finally, viruses may revert to their infective type and cause disease in a patient. Virologists, gene therapists, and biochemists are working to address the shortcomings of viruses by removing or crippling viral genes that cause disease. But the other shortcomings are proving more difficult to address.

Naked DNA

A newer method of gene therapy is naked DNA injection. The gene is synthesized and injected directly into the patient. DNA is not normally found outside the nucleus of human cells, and cells will often respond to the presence of naked DNA by engulfing, or endocytosing, the DNA and incorporating it into their own nuclei. Once there, the wild type gene can become active, correcting the genetic disease. However the rate of DNA incorporation is low and expression of the gene product is poor in most cases. Biochemists and cell biologists are developing methods to enhance incorporation of naked DNA into cell nuclei and to enhance gene product production.

Liposomes

A promising new method of delivering DNA to cells carrying mutant genes is inside liposomes. A liposome is a lipid, or fatty acid, sphere into which DNA or drugs can be inserted for transport in the body. Liposomes can be designed by organic chemists and biochemists with properties that make it more likely that the liposome carrying the DNA will interact with the mutant cells rather than other cells in the body.

Somatic Cell vs. Germ Cell Therapy

In most cases, gene therapy is being used to treat individuals with specific genetic disorders after they are conceived or after birth. In these cases, the gene therapy targets somatic, or body, cells that are the normal cells of everyday function. Treatment of the somatic cells does not prevent the patient from passing the mutated gene on to his or her offspring. However, some gene therapies seek to eliminate the problem from future generations by targeting the germ cells, the sperm and eggs. By providing a corrected copy of the mutant gene to the germ cells, all subsequent generations will possess a corrected copy of the gene and will function normally. A major concern of germ cell therapy is that it opens the door for the creation of so-called designer babies who possess traits selected by the parents and inserted into the babies' genes before conception. Current US law prohibits germ cell therapy.

Future Directions

Gene therapy is still in its infancy. As geneticists learn more about the genes that cause diseases and cancer, they will design better gene therapeutic agents to combat those diseases. Biochemists and organic chemists are working with immunologists to use antibodies to target liposomes and other gene delivery devices to the specific cells that need the wild type gene and to no others. Geneticists and biochemists are working to develop ways to ensure that the wild type gene is correctly inserted into the patient's chromosomes and that it is properly regulated. There is the possibility that if a gene is improperly inserted into the patient's genome the gene therapy could result in a cancerous tumor instead of a cure. Many scientists will be working on improving gene therapy into the foreseeable future.

Summary

Gene therapy involves the insertion of a healthy copy of a gene into the cells of a patient suffering from a genetic disease. Gene therapists possess in depth knowledge of genetics, biochemistry, organic chemistry, mathematics and statistics, and physiology. Gene therapy is still in its infancy and faces a number of problems including gene delivery to appropriate cells, appropriate gene insertion into the chromosomes, and correct regulation of the inserted gene. But advances in virology, biochemistry, immunology, and other scientific fields are addressing many of these shortcomings.

Concept Reinforcement:

1. Which sciences do gene therapists use to develop treatments for patients with genetic diseases?

2. What are the shortcomings of viruses as gene vectors?

3. What is the difference between somatic cell gene therapy and germ cell gene therapy?

Chapter 21 – Human Genome Project

Chapter Objective:

- Understand the significance that science had in the Human Genome Project, and analyze the impact it had on future projects

Introduction

The Human Genome Project was a 13-year project to identify all the genes in human DNA, determine the sequences of human DNA, store the information, improve tools for data analysis, transfer technology to the private sector, and address the ethical, legal, and social issues of the project. Breakthroughs in computer technology, robotics, optical engineering, biochemistry, molecular biology, mathematics and statistics, and genetics allowed the Project to achieve its goal of sequencing the entire human genome by 2002. While the Project itself is over, the data analysis will take years of effort by biostatisticians, bioinformaticists, geneticists, molecular biologists, pharmacogenomicists, and pharmacogeneticists.

Impact of Technology on the Human Genome Project

When the Human Genome Project first began, DNA sequencing was slow compared to today's standards. Researchers at the time knew that if they were to achieve their goal of sequencing the entire human genome by 2002 they would have to develop sequencing technologies that would work at least two to three times faster than the technologies available in 1998 when the Project began. The Project's initial goal was to sequence 250 megabases (Mb, million DNA molecules) per year for less than $0.25 per base. Advances in computer science, robotics, and optical engineering resulted in the ability to sequence over 1,400 Mb per year for under $0.09 per base by 2002. The nearly fourteen-fold increase in speed at a cost nearly one third of the original goal allowed the Human Genome Project to achieve its goal of sequencing the entire human genome by 2002.

Impact of Mathematics and Statistics on the Human Genome Project

The human genome is really a code in the sense that the language of DNA, with DNA bases "read" in triplets called codons, are transcribed, or rewritten into RNA, then translated into an entirely different "language": proteins. For that reason, the DNA sequence can be treated as any other cryptographic code used by intelligence services around the world. Cryptanalysts, or code-breakers, rely on mathematics and statistics to break codes and predict the sequence of letters, numbers, or symbols that will indicate letters, words, phrases, or ideas not yet directly observed. Similar mathematical and statistical analyses were conducted on the human genome sequence to more reliably predict which primers, short segments of DNA used to initiate the sequencing process, should be used. Such primers could also be developed if scientists had even slight knowledge of what a gene or gene product's chemical structure was. The genome could then be screened with the predicted primers to hunt for genes which could then be identified and characterized. The use of computer programming, mathematics and statistics to store, search, and interpret genetic data is now the field of bioinformatics, and biostatistics.

Impact of Chemistry on the Human Genome Project

When the Human Genome Project began, the only way to visualize DNA was to perform gel electrophoresis, a time consuming task that was limited in the number of bases that could be examined at any one time. Each strand of DNA was cut using enzymes and the fragments were separated on a gel matrix by size. Scientists could then deduce the sequence like a puzzle because the cut ends of the DNA fragments would always be the same for a specific enzyme. Current technology uses a different approach. Tagging DNA with fluorescent molecules to allow optical detection with a laser. The optical detector was developed by organic and biochemists as a means to follow DNA replication in cells. Real time sequencing can now be achieved using fluorescence technology in which fluorescently tagged DNA molecules are allowed to pair with, or hybridize with, the DNA to be sequenced. The fluorescent molecule is detected by an optical reader and the sequence can be "read" as the DNA strand grows.

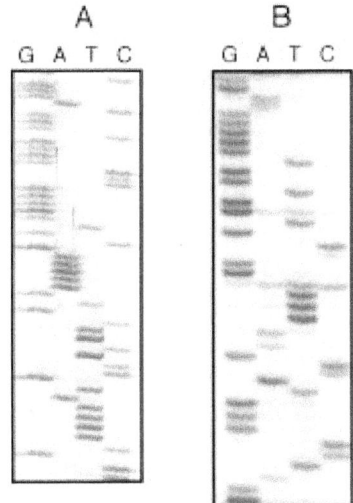

Gel electophoresis of DNA to determine its sequence.

Summary

Breakthroughs in computer technology, robotics, optical engineering, biochemistry, molecular biology, mathematics and statistics, and genetics allowed the Human Genome Project to achieve its goal of sequencing the entire human genome by 2002. While the Project itself is over, the data analysis will take years of effort by biostatisticians, bioinformaticists, geneticists, molecular biologists, pharmacogenomicists, and pharmacogeneticists. Data analysis tools and DNA processing equipment developed during the Project continue to be used and improved as a result of the new scientific discoveries being made in molecular genetics that have been spawned by the results of the Project.

Concept Reinforcement:

1. Which sciences were used to achieve the goals of the Human Genome Project?

2. How did mathematics and statistics advance the goals of the Human Genome Project?

3. How did changes in computer science and robotics help the Human Genome Project achieve its goals?

Chapter 22 – Cloning

Chapter Objective:

- Understand the significance that science has in cloning, and analyze its impact on the future

Introduction

Cloning raises many fears in the public. Science fiction writers have used clones as villains in their novels for decades. But what is a clone? A clone is an organism that is an exact genetic duplicate of another. Could such a thing ever happen in Nature? It happens all the time. Any organism that reproduces asexually is a clone. Bacteria, protozoa, many plants, insects, some fish, and amphibians all are capable of or exclusively reproduce asexually, creating clones of themselves. In the US, one out of every 1,000 births results in identical twins. Their genes are exactly the same and they are clones of one another.

Dolly the sheep and her offspring Polly

How is Cloning Accomplished?

Cloning can be accomplished in a number of ways in lower organisms, and as we have seen, some clone themselves naturally. However, cloning of mammals requires specialized knowledge of molecular biology, reproductive physiology, cell biology, and genetics. Nuclear cloning, first successfully accomplished by Dr. Ian Wilmut in Dolly the sheep, involves the removal of the nucleus of an egg cell, or oocyte, removal of the nucleus of a somatic, or body, cell, and transfer of the somatic nucleus to the oocyte. The newly created cell begins to divide and acts exactly like a fertilized oocyte. It continues to divide and differentiate into a completely new individual.

Cloning in Plants

Cloning has been conducted in plants for centuries. In its simplest form, a portion of a plant is cut free from the rest of the plant and placed in a growth medium to encourage the growth of roots to result in a new plant with the same genes as the one from which it is cut. The technique is used daily in the ornamental plant industry. Plant scientists may add a step in cloning to include gene transfer, in which a gene from a different species is inserted into the plant's genome. The resulting genetically modified plant is cloned and tested to determine whether the desired characteristics have been transferred.

Experimental Cloning

Laboratory mice can be successfully cloned. This results in a large number of genetically identical individuals. One of the shortcomings of many experiments has been the interaction between the genetics of the test animal and the experimental treatment. By cloning experimental animals, scientists can eliminate the genetic variation among the animals. The end result is that any difference seen in the experiment is due to the treatment and not genetics. On the other hand, cloning experimental animals allows scientists to introduce mutations into one and observe the effect of the mutated gene compared to a twin animal that has not been treated. This permits geneticists to observe the effect of specific genes in an organism. More rapid progress in the biomedical fields has been achieved because of cloning.

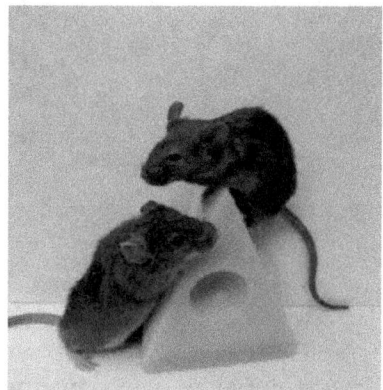

Identical mice. The one on the left was cloned from the one on the right

Cellular Cloning

Individual cells can be cloned and large colonies of them can be developed. This technique is presently used in leukemia treatment. Leukemia is a cancer of the blood cells, usually the white blood cells. Cells from a patient's bone marrow that would normally form blood cells are removed, screened for the absence of the mutant gene, and grown in culture to create clones of a patient's own blood-producing, or hematopoietic, cells. The patient is then subjected to a massive dose of radiation to kill all of the hematopoietic cells in his or her body. The cancerous blood cells are killed along with all of the rest. Then, the patient's own hematopoietic cells that were cloned can be replaced so there is no danger of immunological rejection. Similar cloning techniques can be used to replace damaged tissue and perhaps eventually whole organs from a patient's own cells.

Revival of Extinct Species

If cloning lives up to its promise of creating an exact genetic duplicate of an organism (if the organism's DNA is available), extinct species may be revived. By collecting a sample of DNA from the skin, bone, or other tissue of an extinct animal, it may be possible to insert that DNA into a related species' oocyte and recreate a species that is now lost forever. Combining embryo transfer technology with cloning may achieve this laudable goal. The science fiction novel Jurassic Park by Michael Crichton, was based on this premise.

Reproductive Cloning

Reproductive cloning raises many concerns in the general public. The fear of many copies of an individual and the potential misuse of the technology are of grave concern. However, there are many couples today who are unable to conceive children of their own. Reproductive cloning may provide an answer. A cell nucleus from one of the parents could be placed in the oocyte of the woman and a genetically related offspring could be produced. The ethical concerns raised by reproductive cloning have prevented its legalization.

Summary

A clone is an organism that is an exact genetic duplicate of another. Bacteria, protozoa, many plants, insects, some fish, and amphibians all are capable of creating clones of themselves. In the US, one out of every thousand births results in identical twins and they are clones of one another. Cloning of mammals requires specialized knowledge of molecular biology, reproductive physiology, cell biology, and genetics. Nuclear transfer cloning involves the removal of the nucleus of an oocyte, removal of the nucleus of a somatic cell, and transfer of the somatic nucleus to the oocyte. Cloning can be used for experimental purposes, creation of genetically modified organisms, replacing injured or diseased tissues, revival of extinct species, and reproduction. The ethical questions raised by cloning require great care in the use of the science in society.

Concept Reinforcement:

1. What are the sciences involved in nuclear transfer cloning?

2. How can cloning benefit research scientists conducting experiments?

3. How do scientists create clones?

Chapter 23 – Agriculture Improved Crop Yield

Chapter Objective:

- Understand the significance that science has in agriculture to improve crop yield, and analyze its impact on future crop

Introduction

Science has long played a significant role in American agriculture. Thomas Jefferson, the third President of the United States, had many crop varieties shipped from Europe to his farm in Virginia, where he kept scientific records of their productivity and recommended the best varieties to neighboring planters. But it was not until the Civil War in 1862 that agriculture in the United States was truly scientifically studied in a systematic way.

Abraham Lincoln realized the importance of adequate food supplies to support the Union War effort, and under his administration, Congress passed legislation to create the land grant university system. The states were granted federal lands that they were permitted to sell to finance the creation of universities to study "the agricultural and mechanical arts" or agriculture and engineering as we know them today. Many of the largest universities in the US today are land grant universities including Pennsylvania State University, Michigan State University, Cornell University, Clemson, the Ohio State University, Texas A&M University, the University of Wisconsin, and many more. In fact, every state in the US has at least one land grant university. The scientists at these universities are a large part of the reason US agriculture is as productive as it is today.

How productive is American agriculture? Fewer than 2% of the US population actually lives or works on a farm or ranch. But the US leads the world in food production. The average farm worker in the US feeds 135 people. Ninety of those people live in the US, the other 45 live overseas. In Jefferson's day, the average farm worker fed two other people. This enormous change in agricultural productivity is due in large part to major historical events that drove scientists to find solutions to increasing agricultural output.

The Dust Bowl

Dust storm in the 1930's

From 1930 to 1936, an extensive drought coupled with poor farming techniques lead to the loss of millions of acres of topsoil in the Great Plains. Scientists at land grant universities developed new methods of working the land to reduce soil erosion including contour plowing, strip cropping, use of cover crops and wind breaks, and today a practice called "no-till" in which no plowing is done and seeds are "drilled" into the ground. Although soil erosion continues at a rapid rate today, it is nothing like the loss seen in the 1930's because of improved tillage practices developed by soil and crop scientists.

The Haber Process

In 1918, a German chemist named Fritz Haber won the Nobel Prize for developing a method of "fixing" nitrogen. This process is critical to today's agriculture. Although the atmosphere is 71% nitrogen, most plants and animals cannot use it. Nitrogen is an essential element in proteins. The Haber process allows chemical companies to manufacture nitrogen fertilizer. Nitrogen fertilizer is believed to be responsible for between 30 and 40% of US agricultural productivity. Haber was also important in the development of a number of pesticides, although they are not generally used today.

World Wars I and II

Ironically, Haber is also known as the father of chemical warfare, and was responsible for the development of deadly gas poisons for use in World War I. Much of the agricultural benefit of his work was an offshoot of his more deadly work and not a direct goal. It is even more ironic that his methods were an integral part of the Allied defeat of Germany in World War II. While the German Army had difficulty supplying food to its troops, the Allies had surplus food.

The developments of machines during the war revolutionized the agricultural industry.

The development of tanks and other machines of war also revolutionized American agriculture. Tractors, harvesters, and other farm equipment were improved and their cost reduced as a result of new tank design and added factory capacity. They were considered a critical part of the war effort to provide more food with fewer workers, many of whom were off at war. Mechanical engineers were especially involved in the work and a new agricultural specialty, agricultural engineering, was born from the new industry of mechanized agriculture.

The Green Revolution 1

In the late 1950s through the 1960s, plant breeders and agronomists worked together with governments of developing companies and charitable foundations to create the Green Revolution. Mexico, India, and other developing nations regularly faced famine as a result of poor agricultural yields. Researchers developed new varieties of staple crops, especially rice, which produced five to ten times the yield of native varieties. Combined with fertilizers, pesticides, and irrigation projects, these new varieties produced such bountiful harvests that many nations that had been food importers became exporters. Problems eventually developed because of the indiscriminate use of pesticides and the overuse of fertilizers that curtailed crop productivity. Scientists continue to develop newer and better combinations of cropping systems that are more compatible with developing nations' agricultural practices.

The Green Revolution 2

Today, crop scientists are developing new, genetically engineered crops that are pest resistant, drought tolerant, more productive, and easier to harvest and store. By combining their knowledge of the needs of developing countries' farmers, genetics, and cloning, they can produce crops that are more nutritious, crops that can serve as vaccines against common diseases, and crops that have improved productivity under adverse conditions which are being created for export to farmers in the developing world. Many of these advances are still in the early phases of testing, and there are numerous environmental concerns, including the creation of "superweeds" and genetic pollution. But the needs of many hungry people around the globe are driving development.

Summary

Science has been a factor in agriculture and improved crop yield for as long as people have been farming. From basic selection of the best fruits of the harvest for seeds for the next year to genetic engineering of novel crops, scientists have contributed to feeding the world. American farmers are able to feed 135 people in the US and around the world for every farm worker, up from 2 people in the late 18th century. Without the contributions of science to improved crop yields, almost every citizen of the US would be a farmer, toiling in a field from dawn to dusk just to feed himself and his family. Society as we know it today would be very different were it not for science's contribution to improved crop yield.

Concept Reinforcement:

1. What are some of the major events that contributed to science's involvement in crop production?

2. What scientific advances allow US farm workers to feed 135 other people?

3. What are some of the environmental concerns raised by agricultural cropping practices?

Chapter 24 – Agriculture and Environmental Stress Reduction

Chapter Objective:

- Understand the significance that science has in agriculture to reduce environmental stress

Introduction

Stress is caused by environmental factors that disturb the metabolic equilibrium of the organism. An organism's response to stress is an attempt to return to metabolic equilibrium. Science has shown that a certain level of stress is important to normal growth and development. However, excess stress leads to poor performance, inhibited growth, decreased immune function, and suppression of reproduction. Agricultural systems are designed to reduce environmental stress to crops and animals. At the same time, agricultural practices often increase stress. Farmers and ranchers must have compensatory strategies to alleviate the stresses of their management practices on crops and livestock. Agricultural scientists work to develop methods to reduce environmental and management stress on crops and livestock.

Water

One of the most limiting factors in any agricultural production system is water. Rainfall is often not available when plants need it most, and sometimes rainfall abundance is such that crops drown. Irrigation is expensive, and in some states water regulations limit farmers' and ranchers' use of irrigation water. In response to the need for appropriate amounts of water at the appropriate times, scientists have developed sophisticated irrigation systems that include soil moisture sensors and automated valves and meters that only supply as much water as is needed. Researchers at the University of California discovered that when grape vines are water stressed they emit an ultrasonic sound. In other words, the grape vines "ask" for water. Coupling an ultrasonic detector with an irrigation system's valves allows water to be supplied to the plant only when it "asks" for water, reducing overall water use and runoff.

Nutrition

Plants and animals both require appropriate nutrition to grow and reproduce properly. Normally we don't think of plants needing "nutrition," but plants draw their nutrients from the soil. Soil scientists determine the chemical and physical properties of the soil to determine which nutrients are present, the abundance of those nutrients, and their availability to the plant. In some soils, nutrients are abundant, but not available to the plants because the physical properties of the soil bind the nutrients to the soil particles tightly. Plant scientists determine the nutrient requirements of the crop variety by conducting experiments to determine the conditions under which the plant grows best. Combining the knowledge of the plant's needs with the knowledge of the soil's potential, agronomists make recommendations on tillage and fertilization practices to maximize crop yield.

Nutritionists help to determine the nutritional requirements of agriculturally important species.

Animals, on the other hand, are fed special diets to maximize their productivity. Nutritionists conduct experiments to determine the nutrient requirements of agriculturally important species. They also conduct experiments on potential feeds to determine the nutrients present and available to the animal. Nutritionists make recommendations on animal rations based on the age, size, growth rate, milk or egg production, and stage of the life cycle of the animal to maximize production without undue stress. They inform farmers and ranchers what they should feed their livestock and how much. Horses, for example, can suffer laminitis, or founder, if they eat too much grain. Laminitis is a painful inflammation of the layer of tissue between the hoof and the bone. In severe cases of laminitis the bone can actually rotate so far downward that it punctures the hoof and the horse must be euthanized.

Shelter

Pigs under the hoop building. These buildings are less expensive to put up than confinement buildings. Ventilation is also superior and the pigs are exposed to sunlight.

Many animals are grown in confinement systems. Swine, poultry, and beef cattle are housed in enclosures and fed high energy rations to increase their weight gain and improve the characteristics of the meat. However, overcrowding can lead to stress, which reduces feed consumption, increases aggression, and lowers both rates of gain and meat quality. Behavioral scientists study livestock behavior and propose management practices to reduce stress and aggression. Facilities are designed to provide each animal with an adequate amount of personal space, bedding space, and access to feed. In the case of swine and poultry, temperature and humidity are controlled to prevent overheating. In some instances of natural disasters resulting in power failures in the Midwestern US, large-scale die-offs of poultry and swine have been reported due to the failure of these systems. Lighting in poultry houses is usually dim to reduce aggression. Workers in livestock systems are trained to observe animals for signs of stress. Workers separate these animals from the others and provide supportive treatment to return the animal to health and reduce its stress.

Summary

Stress is caused by environmental factors that disturb metabolic equilibrium. Excess stress leads to poor performance, inhibited growth, decreased immune function, and suppression of reproduction. Agricultural systems are designed by agricultural scientists to reduce environmental stress to crops and animals. Knowledge of the water, nutritional, and shelter requirements of crops and livestock are combined with knowledge of computer science, nutrition, behavioral science, and management science to reduce stress to manageable levels and boost productivity.

Concept Reinforcement:

1. What are some of the sciences used to reduce vulnerability of crops and livestock to environmental stress?

2. How can water availability be controlled?

3. What are the important nutritional considerations to maximize plants' and animals' productivity?

Chapter 25 – Crop Nutritional Quality

Chapter Objective:

- Understand the significance that science has in agriculture to improve crop nutrition and analyze its impact on the future

Introduction

Since the dawn of agriculture, humans have altered the nutritional quality of the crops they produce. Primitive farmers selected crops for sweetness, an indication of sugar content and a much needed nutrient for energy production. In fact, most things that taste good do so because they are nutritionally high in value. Sugar, fat, and protein all taste good to people living a subsistence lifestyle. To those of us who have access to many more calories than are necessary for mere subsistence, some highly nutritious substances are not as tasty because our sense of taste is dulled by overstimulation. In more modern times, chemistry has permitted scientists to determine the exact chemical nutrient composition of nearly everything we eat. Any packaged food in the grocery store has its nutrient content printed on the label. Plant and animal scientists have devised agricultural methods and genetic engineering protocols to improve the nutritional value of plant and animal crops.

Crop Nutritional Value

In many ways, crops' nutritional values are dependent upon the soil in which they are grown. Crops grown in soils lacking iodine, for example, have little iodine in their tissues. People eating those crops must supplement their diets with iodine to avoid developing goiter, a nutritional disease characteristic of iodine deficiency. Soil scientists study the chemical and physical properties of the soil to find ways to either increase the availability of the nutrients already present in the soil, or to supplement the soil with fertilizers to enhance crop nutritional quality.

Golden rice on the right side of the photo is a type of transgenic rice which produces the precursor to Vitamin A.

However, with the advent of genetic engineering and the ability to transfer genes among species, changes in the nutritional value of crops can be manipulated in ways we never could before. For example, rice is a staple of diets in Asia. People in Asia get the majority of their calories each day from rice. Unfortunately, rice is deficient in vitamin A. Without sufficient intake of vitamin A, pregnant women face an increased risk of stillbirth. Between 250,000 and 500,000 children go blind each year due to vitamin A deficiency, and roughly half of them die within a year. Scientists at the Golden Rice Project developed transgenic rice that is capable of producing β-carotene, the precursor to vitamin A, in the grain to combat this problem. Golden rice is yellow instead of the normal white because of its high levels of β-carotene.

Meat Nutritional Value

Little attention was paid to nutrient content of meat products prior to the 1980s. Meat has long been noted for its high nutrient density, more nutrients per calorie consumed, compared to plant-derived foods. High protein, mineral, and vitamin content, especially vitamin B12 which is not available from plants, not to mention taste, assured meat a place on the plate of the vast majority of Americans. However, in the 1970s and 80s the connection between foods high in saturated fat like butter, highly marbled beef, and pork with heart disease, and falling sales raised concerns in the livestock industry about the nutritional quality of their product.

Fig. 1

```
5'TGCGCGAACGACAACAAGGGGGTAAACGGGAAAGATCCCCTTGGGGAGGGGTATTCAGTCAGAAGGTA
ATGGTTTCGAACAAAGACAAACCCTCTACCCCATCAGTGACCTGGAGGCAGTGAGGAGGGGCCAGGCCTG
AGAAATATTTCAGAAGGTAATTTACTTTTCTTTGGCAGGAGGGATTTTAATCTCTTAAAATGAGATTAAG
AAGGAGGGAAGGTTCTAGGTGATCCTGTCTGGTCTGAGTAATTAGGTGAAAGGCAGTGATGTTTGCAGGA
AGAAAAAAAGATGAAAAGAACAGGTCTTGGGGTGATGACTCACGGTGCCAGTTATTCAGCGAGCACTTAG
AGAATTCCCAGTATGTGTTGGATGCTGTTTCAGGGGAGTCATGACTGAGACAGGCAGAGTTCCTGCTCAT
GTGGCGAACAGAATGAAAAATGTAAAGAAGGAATAAGAAGTTTCAATGATAAAGACCAAAATGAAGGATC
CAGGGAGTGATAGAGAATGAAAATGAGAAAGAAGGAACATTCTAGGTCAGAGAAGTCCTCTTGGAGGACA
CACCATGTGAGCTGAGATGTGTACGGTTATCAGCT**CTTCTGTGCTCGTTCGCGCA**3'
```

Part of a gene sequence for tenderness in beef.

Geneticists began searching for genes that control meat quality including tenderness, for which a genetic marker has been found. They also sought ways to reduce fat deposition under the hide and between the muscles without decreasing marbling within the meat. In this way, they sought to reduce overall fat content while maintaining flavor and tenderness. As a result of genetic selection programs designed by geneticists and nutrition programs developed by animal nutritionists initiated by beef and swine producers' organizations, beef and pork are leaner and healthier than ever. New findings in human nutrition and disease research have also debunked some of the bad press animal products have received and meat consumption in the US is on the rise.

Summary

Chemistry has permitted scientists to determine the exact chemical nutrient composition of nearly everything we eat. Any packaged food in the grocery store has its nutrient content printed on the label. Plant and animal scientists have devised agricultural methods and genetic engineering protocols to improve the nutritional value of plant and animal crops. Inserting genes into plants to improve their nutritional quality and prevent nutritional diseases is a growing area of crop research and development. Animal production and selection systems are being changed by knowledge of genetic factors controlling meat quality. As biotechnology and genetic research advance, animal production systems may eventually focus on directly controlling an animal's genes during growth and production.

Concept Reinforcement:

1. What are some of the sciences that are used to improve plant and animal crop nutritional values?

2. How have scientists used genetic engineering to alter the nutritional value of plant foods?

3. How have scientists used genetics to alter meat product's nutritional values?

Chapter 26 – Improved Taste, Texture and Appearance of Crops

Chapter Objective:

- Understand the significance that science has in agriculture to improve crop quality and appearance

Introduction

Consumers in America are very aware of the quality and appearance of the produce they purchase in the store. Under natural conditions, fruits and vegetables are under continuous attack by pests and diseases. Beetles, caterpillars, aphids, leaf miners, mold, fungi, birds, squirrels, mice, deer, and other organisms damage or deface crops making them unsuitable for sale or consumption. Weather plays a significant role in the appearance of some fruits and vegetables. Oranges, for example, will become green again after a period of warm weather. Although they are ripe, they will still be green in appearance. Scientists have developed numerous methods to combat these problems and ensure that the desire of the American consumer to have the most unblemished, perfectly formed, properly colored produce available for purchase.

Insect Damage

Crops are under continuous attack by insects and other arthropod pests. Agricultural chemists have long developed poisons and toxins designed to kill insect pests. However, many of these pesticides are harmful to other organisms in the environment, including humans. Pesticide runoff has resulted in fish kills, endangerment of species such as the bald eagle, and human illness and death.

Codling moth larva damages an apple

To reduce or eliminate these concerns, scientists study the biology of pest species to learn ways to disrupt the life cycle of the pest or to develop pesticides that are more specific to the pest and less harmful to non-target species. For example, many moths use a specific pheromone, or chemical attractant, to attract a mate. Moth larvae, caterpillars, are responsible for much of the damage done to US crops. Entomologists, scientists who study insects, developed an artificial pheromone that fruit and vegetable producers can spray near their crops to confuse moth mating. The result is a dramatic reduction in caterpillar infestations and damage.

The advent of genetically modified crops like bt corn has reduced caterpillar damage beyond expectations. Bt corn and other bt crops produce a poison derived from a bacterium. The gene for the poison was transferred from the bacterium to the crop species, and the crop expresses the poison in every cell. Fortunately, the poison is only effective in caterpillar larvae. It is harmless to other organisms. When a caterpillar ingests a portion of the plant, it dies. Because this happens no matter when the caterpillar eats the plant, the caterpillars are killed while they are very small and their damage is not noticeable. The use of bt crops has also reduced the need to apply insecticides, so beneficial insects survive, and there is less chemical runoff.

Weather

The old saying is that everyone talks about the weather but no one does anything about it. While scientists still cannot control the weather, they can affect ripening and coloring of some fruits and vegetables. For example, oranges may revert to green color even when ripe if they are allowed to remain on the tree during a warm period. Plant scientists discovered that ethylene gas, which is responsible for natural fruit ripening, will change the color of an orange from green to orange. Oranges that were not sold as fresh fruit in the past because of their color are now treated with ethylene gas and shipped to the supermarket.

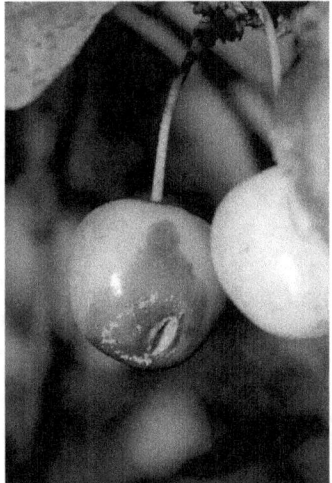

Brown rot on rain-split cherries

Rain frequently comes at times when it is detrimental to the quality of a crop. During periods of heavy rainfall, cherries that are ripe or very nearly so will swell and split the skin, ruining the harvest. Scientists at Michigan State University discovered that spraying a light coating of salty water on the cherries prevented the splitting. They took advantage of a long-known chemical property: osmosis. Water moves from areas with low levels of solutes to areas with high levels of solutes. By putting salt on the surface of the cherries, water from moves from the inside of the fruit through the skin to the surface and the cherry fails to swell and split.

Summary

Agricultural crops come under attack by a variety of pest species that can damage their quality and appearance. Environmental effects such as temperature and precipitation can adversely affect crop quality. Plant scientists, soil scientists, entomologists, chemists, genetic engineers, and others work to reduce the damage caused by plant pests. The high quality of American produce is maintained by the cooperative efforts of scientists and farmers.

Concept Reinforcement:

1. What are some of the sciences that are used to reduce pest damage to crops?

2. How have scientists used chemistry to improve fruit's appearance in the supermarket?

3. How have genetically modified crops reduced pest damage while protecting beneficial insects?

Chapter 27 – Reduced Dependence of Pesticides, Fertilizers

Chapter Objective:

- Understand the significance that science has in agriculture in reducing dependence on pesticides and fertilizers for crops

Introduction

While pesticides and fertilizers in the US contribute to an estimated 30 to 50% of agricultural crop harvest each year, the high level of use of agrichemicals has caused enormous environmental problems. The US and the world cannot afford to lose US agricultural production. Because US food exports provide so much of the world's food supply, famine on a massive scale would likely result. Agricultural scientists have been working on methods to reduce the dependence of American agriculture on pesticides and fertilizers.

Pesticides

Entomologists spend a great deal of their time studying the life cycles and feeding requirements of harmful insect pests. By learning the timing of the needs of pest species, it is possible to break the pest's life cycle and prevent it from causing damage. For example, if a farmer knows a pest requires corn when it hatches, the farmer can plant a different crop in his cornfield in alternate years. When the larvae hatch, there is no corn for them to feed on and they die. Because they die before they reach maturity, they do not reproduce and no larvae hatch the following year in that field.

BT corn

In 1996, the first genetically modified crop was introduced to American farmers. Transgenic corn was created using a gene from the bacterium *Bacillus thuringiensis*, or *bt*. This gene causes the corn to manufacture a protein that is harmless to most species, but in the digestive system of Lepidoptera (butterflies and moths) caterpillars, the protein is digested in such a way as to make it toxic to the caterpillar. When caterpillars feed on plants that express the *bt* gene, they die. Other species, however, are not affected. Because the insecticide is present in all of the plant's tissues, many different kinds of caterpillars that damage crops are targeted. As a result, the need for pesticide spraying of *bt* corn and other *bt* crops is much reduced. Cotton, for example, is sprayed as many as 9 to 12 times in a growing season to control boll worms. With the advent of *bt* cotton, the number of sprayings has been reduced to 3 to 4 times per growing season.

Weeds are another pest that must be controlled on the farm. Herbicides have been used with great success, but they must be applied at very specific stages of a plant's life cycle or they are not effective. Many times, the weed is in the same stage of development as the crop. In order to avoid losing the crop to the herbicide, the farmer may not spray until a less effective stage of the life cycle, or he may have to use a more harmful herbicide. The advent of herbicide resistant crops, so-called Round-up Ready crops, has eliminated this problem. Glyphosate, sold under the name of Round-up, is a relatively safe herbicide that is broken down by soil microorganisms. However, it is a broad spectrum herbicide that will kill crops as well as weeds. When farmers plant Round-up Ready crops, the crop contains a gene from the soil microbe that breaks down the herbicide and prevents the plant's death. The weeds, on the other hand, are killed. Farmers are able to use Round-up at the best time to kill weeds and use fewer applications as a result. The combination of fewer applications with a less dangerous herbicide than would otherwise be used has reduced the environmental presence of more dangerous herbicides.

Not all is rosy in the debate over the use of genetically modified (GM) crops. Many activists and some scientists are concerned about the development of resistance by insects and weeds. Others are concerned about the possibility of unwanted gene transfer into either weedy relatives of crop plants, making them difficult to control, or into organic crops thereby making them non-organic. Agricultural scientists are currently working on recommendations and strategies to prevent or at least reduce the incidence of these unwanted effects. But the use of GM crops remains controversial.

Fertilizers

Fertilizers are a critically important part of the agricultural system that feeds us all. However, fertilizer runoff into the Mississippi River drainage system flows into the Gulf of Mexico and creates a vast dead zone where, because of algae blooms and oxygen deficiency, few fish live. Scientists are currently working on solutions to reduce the need for fertilizers. Organic and soil chemists have developed fertilizers that "stick" to the soil and do not readily wash away when rain or irrigation water enters the soil.

Intercropping sorghum, a grain, and pidgeonpea, a legume.

Clover and alfalfa and other legumes are plants that house bacteria in nodules in their roots. The bacteria are capable of converting atmospheric nitrogen into a form the plant can use. Not only do these plants not need nitrogen fertilizer, they can add nitrogen to the soil for other plants to use. Intercropping can also significantly reduce weed growth and increase profits by allowing two crops to be grown in the same field. By developing intercropping systems using legumes or clovers, agronomists help farmers use less fertilizer and get the same results.

Geneticists and molecular biologists are developing new transgenic crops that will not need nitrogen fertilizer. By isolating the genes in legumes and clovers that allow nitrogen-fixing bacteria to live in their roots, plant geneticists hope to create transgenic crops that can make their own nitrogen fertilizer and not require fertilizer.

Summary

Both high and low tech answers are available to farmers to reduce their dependence on pesticides and fertilizers. Knowledge of plant needs and pest species life cycles developed by scientists have allowed recommendations to be developed for crop rotation and intercropping to reduce the need for both pesticides and fertilizers. New technologies in trangenics are allowing farmers to reduce the number of times they spray their fields for pests. New developments are in the production pipeline to provide transgenic organisms that need less fertilizer as well.

Concept Reinforcement:

1. What are some of the sciences that are used in reducing farmers' dependence on pesticides and fertilizers?

2. Why is it important to reduce farmers' dependence on pesticides and fertilizers?

3. How has genetic and molecular biology contributed to reducing farmers' dependence on pesticides and fertilizers?

Chapter 28 – Production of Novel Substances in Crop Plants

Chapter Objective:

- Understand the significance that science has in the production of novel substances in crop plants

Introduction

Plants have long served human need as a food source, a source of medicine, and building materials. Advances in plant breeding and genetics, molecular biology, gene transfer, and biochemistry have allowed scientists to create plants that are more useful than ever. New genetically modified plants can produce more and healthier food, biopharmaceutical products, and industrial chemicals at lower cost and with less environmental impact than current technologies using petroleum based products.

Healthier Foods

Grains are a staple of diets worldwide. Wheat, rice, corn, barley, millet, and oats provide the majority of calories to people around the world. But grains, while high in energy in the form of starch, are often lacking in other important nutrients. For example, in Asia, 250,000 to 500,000 children become blind every year because of vitamin A deficiency. Scientists at the Golden Rice project have inserted a gene into rice that causes the rice to produce β-carotene, the precursor to vitamin A that is used in the human body. Golden rice should reduce or eliminate vitamin A deficiency and the incidence of blindness in Asia's children.

Team of scientists who inserted genes for heart-healthy omega 3 fatty acids into plants.

Genes to produce omega 3 fatty acids, heart-healthy fatty acids normally only available from fish, have been inserted into plants. Commercial production of the plants is still several years off. But the availability of healthier plant oils is on the near horizon.

Vaccines

In much of the developing world, infant mortality rates exceed 15%. For every 1,000 babies born, 150 will die before they reach their 5th birthday. In the US, only 7 children in 1,000 will die before their 5th birthday. Many of the deaths in the developing world are due to diseases that are preventable. Vaccination programs in the US and the developed world have reduced or eliminated many of the major childhood diseases, a primary reason so few children in the US die in early childhood.

Barriers to vaccination in the developing world include cost and adequate transportation. Many vaccines must be refrigerated, a condition that cannot always be met while transporting vaccines in developing countries. Plant geneticists and pharmacologists are working together to insert genes into common crops that cause the plant to manufacture a vaccine in its edible portions. In this way, people in the developing world can grow food that will also provide their children with protection against deadly child hood diseases.

The Cholera vaccine is available in genetically modified rice.

Industrial Chemicals

Some chemicals that have industrial applications are produced by plants naturally. Unfortunately, the quantity produced in each plant is often too low to be recovered economically. Polyhydroxybutyrate (PHB) is a chemical produced by plants and used in making biodegradable plastics. Scientists have inserted the gene for PHB into plant species that are easier to grow in large quantities which induces the plants to overexpress the gene product resulting in higher levels of PHB in the plant's tissues. Commercialization is still in the future, but this and similar products produced by plants will one day replace many of the products we now synthesize from petroleum.

Future Concerns

The primary concerns raised by activists opposed to the use of genetically modified crops are 1) genetic pollution, and 2) safety of the human food supply. Genetic pollution is the breeding of non-genetically modified, or wild type, plants with genetically modified ones, resulting in the spread of genes into the natural population or organic crops. The safety of the food supply is also an important issue. Introduction of novel substances into crop species may result in an allergic reaction in consumers who eat a product unaware that it now contains an allergen. Allergic reactions can be severe, resulting in hospitalization and death. Scientists are working on ways to address the concerns of activists to prevent the contamination of the environment and the food supply.

Summary

Plants have long served human needs, and advances in plant breeding and genetics, molecular biology, gene transfer, and biochemistry have allowed scientists to create plants that are more useful than ever. New genetically modified plants can produce more and healthier food, biopharmaceutical products, and industrial chemicals at lower cost and with less environmental impact than current technologies using petroleum based products. Providing vaccines to developing countries will improve global health and reduce the economic burden imposed by disease on developing countries. Food plants that contain healthier ingredients will help reduce the incidence of heart disease and obesity in the developed world. Industrial chemicals produced from plant materials are more readily biodegradable and will reduce the cost of disposal. The world is moving away from a petroleum based economy and toward a bioeconomy that will bring solutions to current problems, but will undoubtedly raise problems of its own.

Concept Reinforcement:

1. What are some of the sciences that have contributed to introduction of novel substances into crop plants?

2. What are some of the problems associated with the introduction of novel substances into crop plants?

3. How can the introduction of novel substances into crop plants overcome the barriers to vaccination in the developing world?

Chapter 29 – Biological Engineering

Chapter Objective:

- Understand the significance that science has in biological engineering, and analyze its impact on future technologies in this field

Introduction

Biological engineering, or bioengineering, involves the design and improvement of biological systems to produce desired products. Bioengineers manipulate the conditions under which biological systems live, insert novel genes into organisms, and develop enzyme-based systems to create specific end products for society's utilization. These products are typically food technology and biotechnology products. Bioengineering relies on biochemistry, organic chemistry, cellular and molecular biology, microbiology, fluid mechanics, thermodynamics, and biology to design and build productive biological systems.

Pharmaceutical worker at a bioreactor.

Bioreactors

A bioreactor is a container in which chemical reactions carried out by organisms, typically bacteria or yeast, or biologically derived enzymes are housed. Bioreactors must meet the environmental needs of the organisms or enzymes in order to function. Most bacteria and yeast require a neutral or near neutral pH. Temperature must be maintained within optimum ranges so that enzymatic processes continue and the organisms are not killed. Presence of oxygen must often be regulated carefully. Many useful bacteria are anaerobic, and cannot live in the presence of oxygen above minimal levels. Many aerobic organisms contribute to rot and destroy the reaction products if oxygen is present in too high a quantity. Most bioreactors include agitators, or stirring devices, to keep the organisms, enzymes, and nutrients evenly distributed. Light must often be regulated. Bioreactors that rely on plants require sunlight. Bioreactors that rely on bacteria often exclude ultraviolet light as it is damaging to bacteria. Bioreactors produce pharmaceutical chemicals, such as insulin. Beer is produced in a bioreactor using yeast. Bioreactors are also used to treat sewage by breaking down organic sewage components more rapidly than possible in nature.

Tissue engineering

Tissue engineering involves creating tissues or organs to repair or replace damaged tissues in the body. Biological engineers have developed engineered tissues such as Alloderm which can be used to replace burn victims' skin and protect their bodies from infection. Bioartificial livers are a combination of tissue engineering and bioreactors. Liver cells are cultured inside a chamber, and a patient's blood is passed through the chamber using a catheter. The liver cells in the chamber perform the functions of the patient's damaged liver, clearing toxins and metabolizing food. The liver cells do not come into contact with the patient's immune system, avoiding rejection. The patient's health can be prolonged until a suitable transplant can be found or the patient's own liver can repair itself. Tissue engineers are developing collagen scaffolds upon which cells can be seeded to repair bones and cartilage.

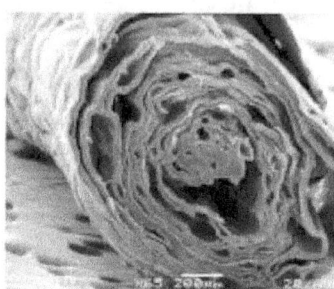

Collagen matrix ready for cell seeding to repair damaged tissue.

Bioremediation

Bioremediation refers to the practice of using organisms to decontaminate the environment. For example, microorganisms have been genetically engineered to "eat" oil. They can be spread on oil spills and will break down the oil into less dangerous byproducts. Highly irrigated soils can become excessively salty, or saline, and crops can no longer be grown on them. Highly valuable farmland is often lost in this way. Tomatoes have been genetically engineered to take up salts from the soil and sequester them in the leaves of the plant, but not the fruit. The tomato is suitable for consumption and the soil salinity is reduced by removing the rest of the plant. Over time, the soil can be returned to its original state of salinity.

Genetically engineered tomatoes can reduce salinity in high saline soils

Summary

Biological engineering manipulates the conditions under which biological systems live, inserts novel genes into organisms, and develops enzyme-based systems to create specific end products for society's utilization. These products are typically food technology and biotechnology products. Bioengineering relies on biochemistry, organic chemistry, cellular and molecular biology, microbiology, fluid mechanics, thermodynamics, and biology to design and build productive biological systems. Biological engineering improves agriculture, medicine, and the environment. Biological engineering is poised to become an important part of the growing bioeconomy in the US.

Concept Reinforcement:

1. What are some of the sciences used in biological engineering?

2. What do biological engineers need to know to design a bioreactor for an organism?

3. How can biological systems be used to clean up the environment?

Chapter 30 – Biodegradation

Chapter Objective:

- Understand the significance that science has in biodegradation, and analyze its impact on the future

Introduction

Biodegradation is the process whereby organic compounds are broken down by enzymes produced by organisms. Biodegradation is a natural process that is responsible for the carbon and nitrogen cycles in the environment. However, scientists have harnessed the power of biodegradation to provide additional benefits to society using their knowledge of microbiology, genetics, cellular and molecular biology, and biochemistry. Power generation and bioremediation are two of the most common applications of biodegradation.

Power Generation

When trash is placed in a landfill, it is densely packed and covered tightly with dirt. The result is an anaerobic environment in which anaerobic bacteria begin to decompose the biodegradable materials in the trash. Byproducts of anaerobic decomposition include methane and hydrogen gas, both of which are readily combustible.

Power plant operating on methane gas from a landfill.

Landfill operators install complex ventilation systems that recover the hydrogen and methane through a series of ducts and pipe the gases to an electrical power station on the landfill site. The gases are burned to generate steam from water that is used to spin turbines that generate electricity. In as little as four years, a landfill operator can recover the cost of building the plant with power sales. Much smaller anaerobic digesters can be installed on dairy farms and other confinement animal operations. These digesters perform the same function as the anaerobic landfill, decomposing the animal waste to create hydrogen and methane that can be used to generate power to run the farm. Surplus power can be sold to local electric companies.

Bioremediation

Many of the chemicals important to modern society are also environmentally toxic. Polychlorinated biphenyls (PCB) were once widely used as refrigerants. PCBs *bioaccumulate*, in other words remain in the tissues and become more concentrated as organisms lower on the food chain are consumed by organisms higher on the food chain. They do not degrade at an appreciable rate in the environment and have spread to every corner of the planet, including Antarctica. For these reasons PCBs were banned in the US and other developed countries. Several bacteria have been discovered that are capable of decomposing PCBs. However, they prefer other chemical substrates as energy sources and will use PCBs only if they must. In natural settings, their rate of PCB degradation is slow. Scientists have identified the enzymes and the genes responsible for the bacteria's ability to degrade PCBs and are attempting to develop methods of using the enzyme rather than the bacteria as the means of breaking PCBs down in the environment.

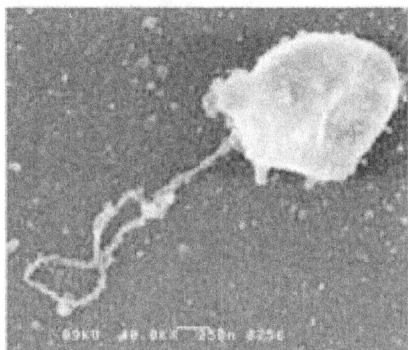

Electron micrograph of a genetically engineered bacterium that degrades PCBs in lab tests.

Summary

Scientists use their knowledge of microbiology, genetics, cellular and molecular biology, and biochemistry to harness biological systems to degrade organic materials and produce products for society. Power generation and a cleaner environment are two of the most common uses for biodegradation today. As advances in molecular biology, genetics, and biochemistry continue to be made, more uses for biodegradation will be created and utilized to improve our lives.

Concept Reinforcement:

1. What are some of the sciences that are used in biodegradation?

2. How is biodegradation used to generate electricity?

3. How can biodegradation be used to improve the environment?

Chapter 31 – Introduction to Information Technology

Chapter Objective:

- Understand the significance of science in information technology

Introduction

Information technology (IT) is the use of computers and computer software to manage, store, process, transmit, retrieve, and manipulate data. Information technology draws heavily on many scientific fields to accomplish this mission. Information technology is also concerned with the security of data and data systems.

Hardware

IT professionals are involved in the care and maintenance of computers and computer systems. Electronic components of computers are complex pieces of engineering. Their complexity leads to malfunctions or hardware failures. Materials science has provided computer manufacturers with new materials that are able to withstand the rigors of the electronic environment.

For example, computer components handle very large electrical currents for their size, and it is relatively easy for microsurges in power to cause elements of the components to short circuit causing physical damage. Knowledge of electronics and the physics of electricity have provided an answer to the problem of surges. Capacitors, electrical storage devices, are installed on computer boards to reduce the likelihood of a surge.

Bulging capacitors caused by voltage spikes can eventually leak and damage nearby components.

Computer components also generate a great deal of heat due to electrical resistance, another aspect of the physics of electricity and energy transfer. Computers come equipped with fans to cool the components. Blockage of the vents can result in heat build-up and occasionally fire. IT professionals must also know how to remove damaged components and replace them. Even static electricity from handling the new components can result in significant damage before they are installed.

Software

Modern software may consist of millions of lines of code, and different software packages are not always compatible with each other. Incompatibilities can result in slow processing or system failure. Many computer users are not knowledgeable about correct installation procedures and may improperly install the software creating problems. Computer science and an understanding of programming are essential to information technologists. IT professionals must be able to install software on computers so that it will function. They must either reconfigure the system's settings, or they must correctly configure the new software package to function within the system.

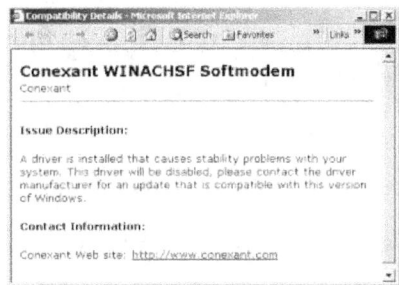

Incompatibility warning for a driver installed on a computer

Security

Attacks by hackers on computer systems cost government agencies and businesses billions of dollars every year. Security software has been developed to reduce the threat, but breaches of security still occur. IT professionals use complex mathematics, including the science of prime numbers and mathematical encryption techniques to reduce or eliminate the threat of sensitive data being accessed by unauthorized users. Even if a system is breached, the information contained within the system is unintelligible without the correct encryption key. Computer scientists specialize in threat detection, data encryption, and intrusion prevention.

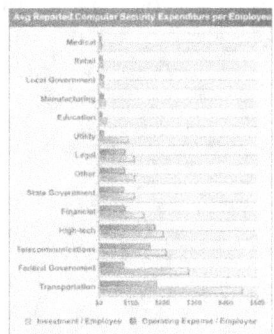

Cost of computer security

System Design

IT professionals advise corporations and government agencies about the hardware and software available to best accomplish the mission of the organization. They must have knowledge of the advances in computer hardware, software, and available transmission systems. They combine this knowledge with their understanding of the compatibility of the hardware and software. IT professionals must understand the needs of the organization so they can develop systems that allow the organization to store, process, transmit, retrieve, and manipulate its data. IT professionals must consider the budget of the company, the sophistication of the users, and the physical limits of the facility's electrical and communications infrastructure. They must finally consider the amount of space available for the system. IT professionals must understand wireless transmission, fiber optic transmission, standard wire transmission, and their capabilities and limitations.

Summary

Information technology is the use of computers and software to manipulate, store, retrieve, transmit, and process data. IT professionals use mathematics, computer science, computer programming, electrical engineering, electronics, and materials science to design, operate, and protect computers and computer systems. Since computer technology changes rapidly, IT professionals must commit to lifelong learning about their profession to provide the best service to their customers.

Concept Reinforcement:

1. What are some of the sciences that are used in information technology?

2. How is mathematics used to increase data security?

3. What are some of the solutions provided by scientists to prevent computer hardware malfunctions?

Chapter 32 – History of Information Technology

Chapter Objective:

- Understand the significance that science has played in the history of information technology, and analyze its impact on future technologies

Introduction

The history of information technology can broadly be considered to encompass the development of writing, simple computational devices like the abacus, and mechanical computing devices. However, this discussion will be confined to electronic computers beginning in the 1940s. Early computers were room-sized machines composed of wires and vacuum tubes costing millions of dollars. They were capable of simple arithmetic calculations and had no data storage capabilities. The calculating power of such machines is now available on credit card sized calculators that are often given away. The advances in information technology have been dramatic, and innovations that test the limits of physics are being developed and tested.

ENIAC

The Electronic Numerical Integrator and Computer (ENIAC) became the first operational electronic computer in 1946. It used vacuum tubes and wires to perform calculations. It could not, however, store its instructions, or software. It was used to calculate ballistic trajectories for the US Army's artillery service. The machine was enormous. Descriptions vary, but ENIAC weighed between 27 and 30 tons and was between 80 and 100 feet long and 8 to 10 feet tall. It was capable of adding, subtracting, multiplying, dividing, and calculating square roots. It was also capable of AND and OR logic functions. The cost of the machine was nearly $500,000.

Electronic Numerical Integrator and Computer

ENIAC and other early computers used punch cards to input data, and vacuum tubes to perform calculations. In 1948 when computers capable of storing their instructions were developed, magnetic drums were used to store instructions internally. Instructions were fed into the machines using punch cards as well.

Transistors

With the discovery of semiconductor materials by materials scientists in the 1940s, electrical engineers were able to design and build transistors. Transistors have electrical properties of insulators, but under certain conditions they can also allow current to flow. This makes them ideal for logic circuits and computation as they can be "opened" or "closed". Transistors replaced vacuum tubes in computers in the late 1950s to early 1960s. Computers became smaller and easier to maintain. They were also significantly faster. However, transistors had to be individually soldered into place by hand to create a circuit, and were bulky compared to modern circuitry.

The first transistor, developed by AT&T Bell Labs in 1947.

Integrated Circuits

In 1958, the first integrated circuits were produced. Integrated circuits differ from transistors in that there are many circuits on a single silicon chip. They are manufactured using photolithography instead of hand construction of each circuit permitting them to be mass produced. Advances in ceramics and other materials sciences allowed the creation of silicon wafers covered with photoreactive substances. When exposed to light, these substances could be easily removed, creating circuits.

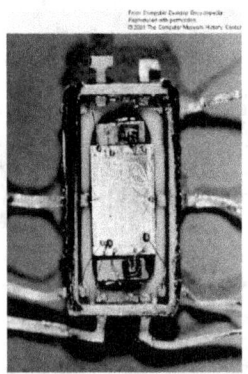

Early integrated circuit

A mask in the shape of the circuit is produced and placed over a light source. In the modern integrated circuit, X-rays are used as the light source because their small wavelength allows for the creation of much smaller circuits. Advances in materials science, the physics of magnetic substances, and electrical engineering made it possible for magnetic tapes and magnetic disks to replace punch cards for programming, data input, and data storage at this time as well.

Early photolithograph ready for reduction

Summary

Numerous advances in electronics, electrical engineering, materials science, mathematics, physics, and chemistry have permitted computers to become increasingly rapid and powerful. Future developments being tested include the use of quantum bits, or qbits that are capable of being in both the on and the off position at the same time and holographic memory systems that store data as small bubbles in plastic that can be read using lasers. As quantum physics and materials science discoveries are made, new applications in computers and information technology will be discovered.

Concept Reinforcement:

1. What are some of the sciences that have contributed to advances in information technology?

2. How did the discovery of semiconducting materials advance computer technology?

3. How did photolithography lead to mass production of integrated circuits?

Chapter 33 – Branches of Information Technology

Chapter Objective:

- Understand the various branches of information technology, and analyze how science has played a role in the branches

Introduction

The information technology (IT) field encompasses every business in the US, from farming to space exploration and from business applications to personal communications with friends and families. No individual can master all of the potential needs of such a diverse array of customers. IT professionals specialize in one of several major branches of information technology. Specializations include programming and software management, computer systems, and database and information management. Each field overlaps with the others, and there are many areas within each specialization. Here we will consider the use of science in the broadest IT specialties.

Programming and Software Management

Computer programmers, software engineers, and computer security specialists all work with the instructions computers use to carry out their calculations and logic functions. Computer programs today can contain millions of lines of code and numerous complex logic loops. IT professionals who work with programs must be firmly grounded in logic, a branch of mathematics. They must be versed in other forms of mathematics as well. Computers are really calculating machines that work with mathematical equations to perform their functions. Computer science and information management play important roles in software development. Computers must receive accurate instructions as to the whereabouts of data and the proper method of retrieving and using data for calculations. Without information management science, programmers would have difficulty accessing stored data.

Computer Systems

Computer systems specialists include computer systems analysts, computer and information systems managers, and computer support specialists. Computer systems specialists design and build networks of computers, routers, data storage devices, and the communications links that connect them into a system. A computer system can be as simple as the one connecting a monitor, keyboard, and printer to a desktop computer, or as complex as the internet itself with its billions of connections and data storage and retrieval systems. Computer systems specialists must also recommend the best software packages to use on the system. Knowledge of computer science, computer engineering, communications, and computer software are indispensible to a computer systems specialist. They use this knowledge to design, build, and maintain computer systems. Computer security specialists are often involved in designing systems to maximize the security of the system while maximizing the ability of authorized users to accomplish their tasks.

Database Administration and Information Management

Database administrators, computer and information research scientists, and computer security specialists are concerned with the storage, preservation, protection, and retrieval of data and information stored on computer systems. Using mathematics, information science and informatics, and advances in materials science they design, build and maintain data storage devices and data storage management systems. They must understand how information can be stored, how it can be lost, how to move information from storage to the user and back without corrupting the information, repairing corrupted data, and encryption techniques to maintain data security. For example, mathematics plays a significant role in database management, especially data security. Data encryption is based on computations using very large prime numbers. Without the correct encryption key, data cannot be retrieved.

Summary

No individual can master all of the potential needs of the diverse array of customers served by information technology. IT professionals specialize in one of several major branches of information technology. Specializations include programming and software management, computer systems, and database and information management. Each field overlaps with the others, and there are many areas within each specialization. Sciences including mathematics, computer science, information science and informatics, logic, systems engineering, and communications technology are just some of the sciences used by the various branches of information technology.

Concept Reinforcement:

1. What are some of the sciences used in information technology?

2. Why is mathematics so important to data and system security?

3. What knowledge is needed to design a computer system?

Chapter 34 – Information Security

Chapter Objective:

- Understand the significance that science has in information security, and analyze its impact on future technologies in this field

Introduction

Federal law defines information security as "protecting information and information systems from unauthorized access, use, disclosure, disruption, modification, or destruction in order to provide integrity... confidentiality ...and availability...". Information technology (IT) professionals use mathematics, including encryption techniques and quantitative risk management, and software engineering to protect sensitive information.

Encryption Technology

Digital encryption systems use algorithms, or equations, and very large numbers to convert data packages into meaningless gibberish if the encryption key is not available. For example, suppose you and a friend wish to communicate secretly, but it is possible that messages will fall into the hands of unintended readers. You and your friend can create a code that shifts each letter down the alphabet by two letters. "See you at nine" would become "uggaqwcvpkpg," a string of letters with no clear meaning unless the interceptor knew the basis of the code or "key."

Encryption systems do the same thing, but with much greater complexity. A hash value is created using an input number and a hashing algorithm. The hashing algorithm serves as the secret you shared with your friend to move each letter down the alphabet by two. The input number is subjected to a mathematical calculation, the hashing algorithm, and an entirely different number, the hash value, is created. Without knowing the algorithm used to create the hash value, it is impossible to recreate the input number. Highly secure encryption systems use 128 bit hash numbers. There are 2^{128} possible combinations for a 128 bit hash number, or 3,402,823,669,209,384,634,633,746,074,300,000,000,000,000,000,000,000,000,000,000,000,000 possible combinations. There is a better chance of winning the lottery every day of your life than accidentally discovering the hash number.

Quantitative Risk Management

IT security specialists use a number of controls to prevent unauthorized access to information. Risk management techniques allow IT security specialists to develop layered defense systems that minimize the risk of unauthorized access and damage.

Physical barriers, doors, locks, key cards, security cameras, and alarms can be used to keep unauthorized individuals away from a system's hardware. Physical attack is one method of damaging an organization's information infrastructure and must be prevented.

A username and password is one of the most commonly used means of protecting information from intruders. Typically, the password is one that involves personal information that would not ordinarily be known to anyone other than the authorized user. For example, your mother's maiden name is a commonly used password identifier. The risk of access by an unauthorized user is minimized by this barrier to access, but authorized users have no trouble accessing needed information.

Level of access can also be controlled. There is no need for an accounting employee to have access to the product information database, so the access of employees can be compartmentalized. Even if an employee's access codes are compromised, the limitation of access to only those parts of the system the employee needs to complete his job limits the damage to the system overall.

Finally, disaster recovery procedures must be put in place in case the worst happens and the system is damaged significantly. Such damage can be physical or due to malware and unauthorized access. Backup storage procedures should be in place and data recovery methods must be ready to retrieve information before the damage is irreversible.

Software Engineering

Antivirus software and firewalls that limit access of individuals outside the system to information contained within the system are commonly used to prevent unauthorized access. Firewalls and antivirus software prevent viruses and other malware, programs designed to damage systems or corrupt data, out of the computer system.

Summary

Information security is the protection of information and information systems from unauthorized access, use, disclosure, disruption, modification, or destruction. Information technology (IT) professionals use mathematics, including encryption techniques and quantitative risk management, and software engineering to protect sensitive information. Physical barriers, electronic barriers, backup systems, and detection systems play an important role in maintaining the security of sensitive information.

Concept Reinforcement:

1. What are some of the sciences used to strengthen information security?

2. How does mathematics contribute to encryption technology to protect data?

3. Describe risk reduction methods that IT security professionals use to ensure data is protected.

Chapter 35 – Networking

Chapter Objective:

- Understand the significance that science has in networking, and analyze its impact on future technologies in this field

Introduction

Computer networking involves communications among computers, servers, routers, printers and other devices, and the protocols that allow communication. Computer networking is very closely related to telecommunications, and many telecommunications applications are also used in computer networking. Computer science, computer engineering, electrical engineering and electronics are just a few of the scientific and engineering disciplines used in computer networking.

LAN, WAN, WLAN, and WWAN

Computer networks are merely computers that are connected in some way so that they can communicate with one another. Local Area Networks, or LANs, are networks of computers spread over a small geographic area. The LAN may be as small as a small suite of offices or as large as a city. Wide Area Networks (WANs) are computer networks that cover a large geographic region, multiple cities or wider to include international networks. LANs and WANs are connected by landlines, either twisted wire (standard phone lines), coaxial cable, or fiberoptic cable.

Creating a LAN or WAN requires an understanding of the data transmission capabilities of the landline connecting the various computers together. Knowledge of electrical engineering allows network designers to determine capacity of the system. Data transmission techniques developed by computer scientists permit not only the safe and secure transmission of data, but allow it to be divided into discreet packets that can be routed along the most rapid and efficient routes and reassembled when they reach their destination. Sometimes signals can be sent over long distances by fiberoptic cable much more quickly than they can be over the "last mile." The "last mile" refers to the landline leading from the last signal booster to the location where the data will be used. The majority of "last mile" landlines are older twisted wire cables and do not transmit as much data as rapidly as fiberoptic cable. Data must be carefully managed at this point by a variety of electronic components to ensure that it is not lost.

WLANs and WWANs are **W**ireless networks. Wireless networks use radio signals to transmit data from computers to radio transmitters that broadcast the signal over distances of up to several miles. For longer transmission distances on wireless networks, satellite relays are used. Because the signal is broadcast, it is possible for unauthorized users to intercept the signal and obtain sensitive data. Network designers must be aware of this possibility and put appropriate security measures in place. Occasionally, the number of users in a wireless "hotspot" exceeds the capacity of the system to transmit data and connectivity can be slowed or lost. Wireless networks are also significantly slower than landline networks. Network engineers are working to develop new methods to route data over wireless networks to reduce or eliminate the slower transmission rate.

Routing Protocols

Routing protocols are instructions to routers. Routers are hubs on a system that determine where to send a data packet. When a data packet arrives at a router, the router determines which of the paths to which it is connected is open, most efficient, and fastest. The data packet is then sent along that route. Network designers must place enough routers with appropriate protocols in place to ensure that data is not lost and that network bottlenecks do not form. Data packets are tagged with addresses and frequently encrypted. Network designers must understand information science, data management, electronics, and cryptography to ensure data arrives safely and uncorrupted at its destination. No small task considering the billions of web pages in existence today.

Summary

Computer networking involves communications among computers, servers, routers, printers and other devices, and the protocols that allow communication. Networks can be very small with a few computers sharing all resources, or they can be as large as the internet. They can transmit information over landlines, wirelessly, or both. Computer science, computer engineering, cryptography and computer security, electrical engineering and electronics are just a few of the scientific and engineering disciplines used in computer networking.

Concept Reinforcement:

1. What are some of the sciences that are used in computer networking?

2. How can computers be connected to form a network?

3. What are routers and why are they important in computer networks?

Chapter 36 – World Wide Web

Chapter Objective:

- Understand the significance that science has in the world wide web, and analyze its impact on future technologies in this field

Introduction

The World Wide Web was created in 1989 by Tim Berners-Lee, an English physicist turned computer scientist, while he was working at CERN, the European particle physics laboratory in Switzerland. He is the Director of the W3C, or World Wide Web Consortium. In that short time, the Web has become such a ubiquitous part of our lives that we hardly notice it unless it is unavailable. Numerous computer scientists, systems engineers, physicists, electrical engineers, mathematicians, and information scientists have created new hardware and applications software that have allowed the Web to grow from "the Cube" in which it was originally housed in 1990 to a truly world wide distributed network of information accessible to anyone with a computer and telecommunications link.

The World Wide Web is not the same as the Internet. The Web is the collection of interconnected documents and computer applications linked by URLs. The Internet is the hardware that houses and transmits the documents that make up the Web. The Web is only one of the services available on the Internet, email being an example of another Internet service.

Information Science

The first Web was housed on a NeXT Cube desktop computer at CERN. Since that time, the Web has grown to include nearly 60 billion indexed pages. Finding a single page out of nearly 60 billion pages housed on servers all over the globe would be impossible were it not for information science. Information science uses mathematics, computer science, and social science to study the ways in which people interact with information and creates addresses for information so that it may be more easily stored and retrieved. Every page on the Web has a Uniform Resource Identifier (URI), commonly referred to as a URL or Uniform Resource Locator. They are not quite the same. The URI can be either a resource location, URL, or resource name, URN. The URN is like someone's name. While the person's name may be known, the person's whereabouts may not be known. The URL, on the other hand, is an address. The location is known, but the names of the occupants may not be known. The W3C and the Internet Engineering Task Force (IEFT) hold regular meetings to create standards for URIs. These standards are developed by computer and information scientists to ensure information can be found on the Web.

The NeXT Cube computer that originally housed the Web in 1990 at CERN.

Programming

Computer scientists and programmers have created new languages for use with the Web. The original internet was almost exclusively text, with few images because of the size of the image file and the slow speed of the hardware available at the time. HTML, ECMAScript, Flash, and Java Script are examples of languages that have been created for use with the Web. HTML allows documents to be created and uploaded to the Web in a format that can be retrieved by an Internet browser. Pages are usually retrieved in a specific order. Text is retrieved immediately, followed by pictures, videos, and applications. ECMAScript, Flash, and Java Script allow programmers to create applications that run on the webpage. When the page is opened, a program in the background runs an application that allows the user to interact with the webpage. Online gaming depends heavily on such applications.

Hardware

Although the Web is a collection of interconnected documents, it would not play the important role it does in daily life were it not for advances in computer design, telecommunications devices, and routers. Computer engineers, physicists, materials scientists, and mathematicians have developed important breakthroughs in computer speed and sophistication that have allowed the development of increased complexity and richness of Web content. Although often referred to as the World Wide Wait, the Web accomplishes amazing feats of document retrieval and storage due in large measure to advances in hardware design.

Future Directions

The Web is still in its infancy, having only existed for the past 18 years. Advances in physics and mathematics are driving design changes in data storage devices that will increase the speed and accuracy of data retrieval. New developments in programming languages and techniques will make Web applications richer and more interactive. Advances in mathematics and encryption technology will improve computer security and at the same time allow for more rapid data transmission. Information scientists are developing more intuitive methods of identifying and storing data to make retrieval easier and faster. The World Wide Web has changed dramatically since it was first housed on a single desktop computer in 1990. With the exponential increase in the number of users and the innovations they can bring to the Web, the future of the Web is limitless.

Summary

Since 1989, the Web has become such a ubiquitous part of our lives that we hardly notice it unless it is unavailable. Numerous computer scientists, systems engineers, physicists, electrical engineers, mathematicians, and information scientists have created new data storage and retrieval standards, more powerful programming languages, and faster hardware that have allowed the Web to grow from "the Cube" in which it was originally housed in 1990 to a truly world wide distributed network of information accessible to anyone with a computer and telecommunications link.

Concept Reinforcement:

1. What are some of the sciences that have allowed the Web to become the global interconnected collection of documents it has become today?

2. Why is information science so important to the continued growth of the Web?

3. How is science contributing to the future development of the Web?

Chapter 37 – Local Area Network

Chapter Objective:

- Understand the significance that science has in local area networks, and analyze its impact on future technologies in this field

Introduction

A Local Area Network (LAN) is an interconnected group of computers capable of sharing files and hardware resources such as printers. LANs cover small geographic areas, an office, a school, a group of buildings, perhaps even a city (Municipal Area Network or MAN). They have very high speed data transfer rates and usually use communications lines owned by the network. Computer scientists, computer engineers, software engineers, mathematicians, and electronics professionals contribute to the knowledge base that allows computers in a LAN to communicate securely.

Hardware

Computer networks would not be available were it not possible to link them together in some physical way. LANs can operate either over a wired or a wireless connection. Wired connections can use phone lines, DSL cable, or Ethernet cables. Wires run from the connected computers to a switch, hub, or router. These devices decide where to send data packets based on the addresses attached to the data packet. Wireless LANs transmit radio signals from computer to computer. Each computer in the network must have a transmitter/receiver capable of linking to the wireless router. The wireless router directs data in the same way a wired router does. Electrical engineers, physicists, and materials scientists develop the switches, hubs, routers, wireless devices, and wires that make wireless and wired networks possible. As new materials become available, faster and more accurate routers are being developed. Some of the new technology being tested pushes the boundaries of physics disciplines including optics and quantum physics.

Routers capable of supporting a LAN

Software

Computers sharing a LAN must be able to communicate with one another. Data from multiple machines must travel over the same lines and through the same switches. Data on one machine must be located and retrieved by another. Computer scientists and programmers, along with information scientists have developed programs that allow computers to interact with one another without interfering with each other and the overall system. For example, data to be transmitted is broken down into discreet data packets. Each packet is labeled with information regarding its origin, length, and content. The packet is then sent to the router which detects system use in its vicinity and routes the data packet along open connections to avoid data jams at busy switches. When the data packets arrive at the computer, they do not always arrive in the order in which they were sent. The computer reads the packet label and assembles the packet with the rest of the data in the proper order so it can be displayed on the computer's screen.

Information scientists design protocols for developing labels so that all computers can use the same labeling system. Computer systems engineers and software engineers design software packages that instruct computers how to address and reassemble data packets, and instruct routers how to handle them.

Security

Wireless LANs, in particular, are subject to unauthorized access. Computer thieves have used procedures as simple as driving up to a store, turning on their wireless network detector on their laptop, and intercepting unencrypted credit card numbers from the store's wireless network, connecting the credit card readers at the checkouts to the store's server. Computer scientists, mathematicians, and cryptography professionals design security software to prevent data from being transmitted in an intelligible fashion unless the recipient has the correct decryption key. They design barriers such as username and password systems to keep unauthorized users from accessing the LAN.

Summary

Local Area Networks are interconnected groups of computers capable of sharing files and hardware resources such as printers. LANs cover small geographic areas, have very high speed data transfer rates, and usually use communications lines owned by the network. Computer scientists, computer engineers, software engineers, mathematicians, and electronics professionals contribute to the knowledge base that allows computers in a LAN to communicate securely. Switches, hubs, and routers send data packets over wired or wireless networks by the fastest possible route. Software labels data and instructs computers how to disassemble and reassemble data packets. Security professionals work with mathematicians to develop better encryption protocols to protect LANs from attacks by unauthorized users and malware.

Concept Reinforcement:

1. What are some of the sciences used to design and improve LANs?

2. How has science contributed to the development of routers, switches and hubs?

3. Why is mathematics so important to LAN security?

Chapter 38 – Networking Hardware

Chapter Objective:

- Understand the significance that science has in networking hardware, and analyze its impact on future technologies in this field

Introduction

Networking hardware includes devices such as database and file servers, storage devices, routers, switches, hubs, fiber optic cable, coaxial cable, phone lines, access points, and network interface cards. Advances in physics, materials science, mathematics, computer science, electronics, electrical engineering, and computer science have increased network speed and reliability exponentially since their introduction. New developments in one technology often drive the development of new developments in another technology.

Servers and Storage Devices

Database servers are computers that are optimized for data processing. They carry out the data processing functions on files destined for a network computer and send only the final result to the computer. Advances in physics and computer engineering have dramatically increased the speed of servers. New photo-resistant materials are made into smaller, faster computer circuits. Advances in optics has allowed the development of X-ray lasers, capable of creating circuit images as small as a few nanometers (one billionth, or 10^{-9} meters). Such tiny circuits are beginning to push the boundaries of theoretical quantum physics. Electrons begin to act as particles and less like electrons in a chemical reaction. Quantum states of existence begin to appear in which a gate can be both open and closed at the same time.

Stack of servers on a Local Area Network.

File servers house files and optimize their retrieval for processing by the database server or router to send the file to the computer requesting the file. Advances in mathematics and information science are responsible for improvements in the amount of data that can be stored on a file server. They have also increased the speed at which requests for files are processed. Mathematical models have been developed to predict peak demand periods and routes to transmit data at high speed during times of peak demand. Data address labels are developed by information scientists to allow programs to locate files and retrieve them no matter where in the network they may be.

Routers, Switches, and Hubs

Routers, switches, and hubs direct data traffic along the wires or radio frequencies that are open and most efficient so that data arrives at its destination undegraded. Routers determine data priority settings and direct high priority data onto the fastest connections. The development of high speed switches from new materials, which may soon include superconducting materials, has been accelerated by new knowledge in physics, electronics, and materials science. Advances in wireless communication technology have allowed the development of wireless Internet access in laptop computers.

Wireless router

Transmission Media

In the case of wired connections, twisted wire (phone lines), coaxial cable, and fiber optic cable are available for data transmission. Each of these represents a generational step in data transmission. Phone lines were used by early modems, and are still used to transmit data over the Internet because of their ubiquity. They are slow and not able to carry large amounts of data. New developments in computer science and information science, coupled with new understanding of the physics of electrical transmission over twisted wire lines has allowed data to be split into discrete packets and transmitted more rapidly than though it possible on twisted wire. Fiber optic cable, once thought to be so fast that data transmission rates would not exceed its capabilities is running up against the development of a 10 gigabyte Ethernet connection that is starting to overwhelm its current capabilities. Scientific investigations in optics and the physics of light are being conducted to enhance the fidelity with which fiber optic cables can carry higher speed data transmissions. Wireless systems are still comparatively slow, and easily overwhelmed by high numbers of users. New technologies being developed by physicists and mathematicians to more efficiently use the radio frequencies available to increase data transmission speeds are being developed in the lab.

Fiber optic cable uses light to transmit information.

Summary

Networking hardware includes devices such as database and file servers, storage devices, routers, switches, hubs, fiber optic cable, coaxial cable, phone lines, access points, and network interface cards. Developments in one technology often drive the development of new developments in another technology. As switching speeds increase, cable transmission speeds must also increase. As computer processing speeds increase, switching speeds must increase. Advances in physics, materials science, mathematics, computer science, electronics, electrical engineering, and computer science have increased network speed and reliability.

Concept Reinforcement:

1. What are some of the sciences used in networking hardware operation and development?

2. What new materials are being used in networking hardware operation and development?

3. How do developments in one hardware technology drive changes in another hardware technology?

Chapter 39 – Data Management

Chapter Objective:

- Understand the significance that science has in data management, and analyze its impact on future technologies in this field

Introduction

Data Management Association International defines data management as incorporating data architecture, analysis and design, database administration, data security management, metadata management, data warehousing and business intelligence, reference and master data management, data quality improvement, and unstructured data management to improve data stewardship, strategy, and governance. Data managers rely on advances in information science, computer science, physics, mathematics, and statistics for advances in data management.

Data management body of knowledge

Computer Science

Data managers rely on specialized data management programming language to store, access, and structure data in databases. Structured Query Language (SQL) was developed to access information. Questions (queries) are asked by the programmer in a structured manner. For example, a structured query to locate a building would start with country, then state or province, then city, then street, then street number. Address labels for data are similarly structured. Computer scientists, mathematicians, and information scientists develop new programming languages that are capable of accessing, retrieving, and storing data more efficiently as demand for data services grows. Mathematicians and statisticians design equations to predict the behavior of complex data management systems to identify problems before they occur or propose solutions to existing problems.

Physics and Materials Science

Advances in materials science and optics have allowed the first prototypes of holographic memory devices to be designed and built. Current memory storage devices store data as magnetic or optical bits on their surfaces only. Holographic memory storage devices store data in all three dimensions. Encoded bits are read using laser light. More than one bit of information can be stored in a single location. By shifting the angle of the light used to read the stored data, multiple meanings can be derived from a single point. Current technologies only read a single bit at a time, Holographic memories could be read in parallel so that millions of bits could be read at the same time.

Quantum computers that take advantage of the oddities of quantum physics are also being developed. Normal bits are either 0 or 1. Quantum particles can be in two states at the same time, superposition. A quantum bit, or *qubit*, can be a 0, 1 or in superposition. A pair of qubits can be in 4 possible states, and three qubits can have 8 possible configurations. The number of positions a quantum storage device can have is 2^n, with n being the number of qubits, *at the same time*. A standard computer can have only *one* configuration at a time. By taking advantage of this phenomenon, physicists and computer scientists hope to build data storage machines that are capable of storing greater quantities of data more reliably and with greater accessibility than ever before.

Future Directions

Data management is concerned about the growing number of users requesting data at any one time. Coupled with the explosive growth of data available to request, data storage and retrieval slow unacceptably. New programming languages are being developed by computer scientists to address this problem. Developing new methods of mining data to help managers use information in decision-making is another important area of growth. Again, the wealth of available data sometimes makes it difficult to find the relevant data or even be aware that it exists.

New types of data being stored create additional burdens on the system. In the past, most data was text based. In the future, more audio and video data will be stored. Searching databases for pictures or sounds is not yet possible, but there are a few prototype search engines under development that are designed to find images. New data storage devices that utilize holographic or quantum physics will require additional changes in programming languages. Information science will have to devise new methods of labeling data stored in these new devices. Data management is poised to grow at an explosive rate in the near future as advances in these and other sciences make data more available and accessible.

Summary

Data management is poised to grow exponentially. Data management involves data architecture, analysis and design, database administration, data security management, metadata management, data warehousing and business intelligence, reference and master data management, data quality improvement, and unstructured data management to improve data stewardship, strategy, and governance. Data managers rely on advances in information science, computer science, physics, mathematics, and statistics for advances in data management. New programming languages for new data storage devices will revolutionize data management. But they will place additional burdens on the field as data capacity becomes so large that knowing what data exists becomes a management problem in and of itself.

Concept Reinforcement:

1. What are some of the sciences that are used in data management?

2. How have changes in computer science advanced data management?

3. How are changes in physics and materials science changing data management?

Chapter 40 – Data Storage

Chapter Objective:

- Understand the significance that science has in data storage, and analyze its impact on future technologies in this field

Introduction

Data storage devices preserve information in many formats. Data storage can be as simple as a handwritten note or as complex as the latest in quantum computing prototype qubits. This discussion will be confined to data storage devices associated with electronic devices such as computers, MP3 players, flash drives, and other forms of digital data storage. Data storage has evolved significantly over the past half century. Physicists, mathematicians, materials scientists, computer scientists, and computer engineers have integrated and applied their discoveries to advance data storage from paper punch cards to magnetic spools and tapes, to CDs, DVDs, and flash drives. New discoveries in physics, information science, and materials science are driving changes in the future of data storage.

Hard Drives

Hard drives are the data storage device of choice for most home computer users. Hard drives are magnetic discs that rotate at high speeds where electrical data inputs are converted into magnetic form. Particles are aligned according to their magnetic state as either ones or zeros. As the read/write heads pass over the disc, the magnetic bits are read or encoded, depending upon whether data is being accessed or stored. Advances in materials science have created discs that are able to store more data in less space due to improved magnetic qualities. More sensitive read/write heads have been developed using advances in physics and electronics. Hard drive storage capacity is expected to continue to grow rapidly in the foreseeable future.

Hard drive showing magnetic disc and reader arm.

Optical Media

Optical storage devices include CDs and DVDs. CDs and DVDs are essentially aluminum sandwiched between two layers of translucent plastic. Information is stored on the aluminum surface by creating small pits or leaving a smooth surface called a *land*. As the disc spins in the reader, a laser shines through the plastic and is reflected from the surface of the aluminum. The change in the intensity of the reflected laser light when it shines on a pit is read by an optical sensor and converted into either a zero or a one.

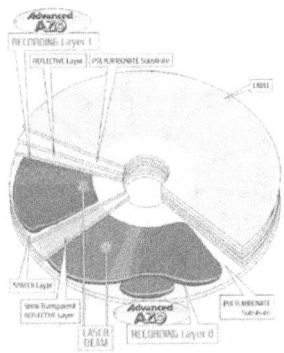

Parts of a dual layer DVD

CD and DVD readers now function at many times their original speed due to faster reader technology. Advances in the physics of lasers and optics are replacing DVD technology with Blu-ray DVD. In Blu-ray technology, the red light laser has been replaced with a blue light laser. Because blue light has a smaller wavelength, blue lasers can read much smaller pits and lands. This allows more data to be stored in the same space.

Flash Drives

Flash drives are essentially upgraded versions of an older technology. Erasable Programmable Read Only Memory (EPROM) chips have been in existence for many years and are commonly found in automotive engines. However, they were slow and had limited data storage capacity. Advances in computer chip technology including new photo-resistant materials, more advanced lasers with smaller wavelengths, and new applications of physics to overcome the quantum effects that occur at such small dimensions have permitted the development of the flash drive. Data can be written directly to the microchips and accessed by the data processing device. Flash drives are rapidly replacing many older data storage devices. They can also be found in MP3 players and other small electronics devices where large data storage needs are coupled with small size. A prototype laptop computer has been developed that weighs less than 2 pounds because it forgoes standard hard drive memory and uses flash memory instead.

Flash drives are small and can be utilitarian or decorative as is this pendant flash drive.

Future Directions

Advances in materials science and optics have allowed the first prototypes of holographic memory devices to be designed and built. Current memory storage devices store data as magnetic or optical bits on their surfaces only. Holographic memory storage devices store data in all three dimensions. Encoded bits are read using laser light. More than one bit of information can be stored in a single location. By shifting the angle of the light used to read the stored data, multiple meanings can be derived from a single point. Current technologies only read a single bit at a time. Holographic memories could be read in parallel so that millions of bits could be read at the same time.

Quantum computers that take advantage of the oddities of quantum physics are also being developed. Normal bits are either 0 or 1. Quantum particles can be in two states at the same time, superposition. A quantum bit, or *qubit*, can be a 0, 1 or in superposition. A pair of qubits can be in 4 possible states, and three qubits can have 8 possible configurations. The number of positions a quantum storage device can have is 2^n, with n being the number of qubits, *at the same time*. A standard computer can have only *one* configuration at a time. By taking advantage of this phenomenon, physicists and computer scientists hope to build data storage machines that are capable of storing greater quantities of data more reliably and more accessible than ever before.

Summary

Data storage has evolved significantly over the past half century. Physicists, mathematicians, materials scientists, computer scientists, and computer engineers have integrated and applied their discoveries to advance data storage from paper punch cards to magnetic spools and tapes, to CDs, DVDs, and flash drives. New discoveries in physics, information science, and materials science are driving changes in the future of data storage. Holographic and quantum data storage methods are on the horizon, increasing data storage capacity and reliability. Data storage advances are driving changes in other fields as the volume of data available makes finding relevant or important data more difficult.

Concept Reinforcement:

1. What are some of the sciences that are used in data storage?

2. How have changes in physics advanced data storage?

3. How are changes in materials science changing data storage?

Chapter 41 – Programming Languages

Chapter Objective:

- Understand the significance that science has in programming languages, and analyze its impact on future technologies in this field

Introduction

The invention of computers created a problem never seen before. How does a human hold a conversation with a machine and instruct the machine how to perform and what to perform? The result was the development of programming languages. The first programming languages were machine languages. Written in the mathematical base 2, they instructed computers directly which switch to turn on (1) and which one to turn off (0). Modern computer languages are much more intuitive for humans to use with recognizable words and phrases. They are also much more powerful, allowing programmers to create complex activities that take up much less memory and processing time.

Mathematics

Few people think of mathematics as a science, but mathematics is the science of quantity, structure, space, and change. Mathematics uses logic and reason to derive its conclusions, and in fact, logic is a mathematical field of study. Mathematics can also be thought of as a language that expresses functions. Computer programming is completely dependent upon mathematics to translate standard spoken and written languages into language a computer can understand.

All computer languages trace their ancestry to the lambda (λ) calculus. Lambda calculus is a method of modeling computations that can be used to develop algorithms that can be understood by a machine. Machines cannot "count" the way humans do. Machines only have on or off, so numbers as people understand them are not comprehensible to a machine unless they are expressed in base 2 where 0 is off and 1 is on. No numbers need be used in the λ calculus, only functions. Numbers can be defined as functions and machines are capable of following instructions based on switching components on or off.

Without λ calculus, computer programming would be very different and much more difficult than it is today. Mathematics, including calculus, is so important to computer programming that whole new kinds of calculus are being developed specifically to address problems in computer programming. The Opus project is one such project which aims to develop a method of creating a model for data memory and data that is independent of the language in which it was written. In this way, even if a computer data storage system changes or the programming language changes, the data will still be readily available. Other calculi include the Calculus of Communicating Systems and the Communicating Sequential Processes Model.

Linguistics

Linguistics is the study of languages and how they are structured, their grammar. The Chomsky hierarchy describes the levels of complexity of language structures and is widely used in the development of computer programming languages. In order for a machine or person to understand an instruction, the words of the instruction must be presented in an order that makes sense based on the structure of the language. "Would you please open the door?" is clearly understood by an English speaker, but "Door would the open please you?" makes no sense. While a human can decipher the true meaning of such nonsense, a machine will follow the instruction to the letter and erroneous results will follow.

Development of a programming language requires computer scientists to create a grammar for the language. The grammar is built according to the Chomsky hierarchy from the simplest of commands and structures to the most abstract. Without the most basic grammar, the more complex grammar will not be understood by the computer.

Object Oriented Languages

Computer scientists and programmers ran into problems with standard programming languages, which are procedural languages or sequential steps of instructions. Procedural languages were not able to adequately compute ships' paths in shipping lanes on the ocean. Programmers needed a language that could allow parts of the program to interact and communicate with one another. As a result, object oriented programming languages were developed. Object oriented languages are especially useful in creating simulations. Modern computer gaming is heavily reliant on object oriented languages.

Related Topics

Compiler theory is the study of writing translators of programming languages from one language to another. Type theory is the theory of classification of programming languages based on the absence of programmable behaviors based on the syntax of the language, i.e., the rules of sentence construction in a language. This is important for programmers to understand so that they do not select a language that is not capable of creating the outcomes they desire. Domain specific languages are sometimes created to address specific problems that are well-defined in a particular domain.

Each of these and other related fields drive changes in programming languages for computers. Advances in computer technology as well as increasingly complex problems that require solutions will drive future developments in programming languages. For example, quantum computer prototypes now being tested will operate on principles far different from those of today's computers and will require new languages to provide instructions while taking quantum effects into consideration.

Summary

Computer programming depends heavily on mathematics as a means of communicating with machines. Modern computer languages are much more intuitive for humans to use with recognizable words and phrases. They are also much more powerful, allowing programmers to create complex activities that take up much less memory and processing time. Computer scientists, mathematicians, and linguists work together to expand theories in their respective disciplines and at the same time propose solutions to problems in computer programming via new programming languages.

Concept Reinforcement:

1. What are some of the sciences used in developing programming languages?

2. Why is mathematics so important to programming languages?

3. Why are computers so literal in their interpretation of instructions?

Chapter 42 – Cryptography

Chapter Objective:

- Understand the significance that science has in cryptography, and analyze its impact on future technologies in this field

Introduction

Cryptography is the study and practice of writing in secret code or hiding information. Cryptography uses advances in mathematics, computer science, and information theory to improve codes and authentication methods for improved security.

Public Key Encryption

Most computer users who transmit secure information use public encryption keys. The public key encryption process is a two-part process. Each user possesses a private encryption key known only to them. The public encryption key is your key that is available to anyone who wants to send you secure information. The sender uses your public encryption key in combination with his private key to encrypt the message. When the message arrives, your computer uses your private encryption key to decode the message.

Secure encryption using the public key method is dependent upon two mathematical and computational concepts. Computationally, the encryption keys are based on very large prime numbers and calculations performed using them. Because the numbers are so large, computationally it is virtually impossible to break an encryption key. Mathematically, the function used to compute the so-called hash number, which is used to encrypt the data is a one-way function. A one-way function is a function that is easy to compute but hard to invert. In other words, the answer to the problem does not readily lend itself to determining how the answer was derived from the initial numbers. Since the basis for public key encryption is that the hash number cannot be used to generate the original data without the key, if the function is not a one-way function or if one-way functions do not exist, the encryption key can be broken. Whether or not one-way functions truly exist is an active area of mathematical research.

Authentication of Identification

The public key portion of the public key encryption method is usually maintained by a trusted intermediary as a *Certificate Authority*. The user's public key is kept on file by the Certificate Authority and its relationship to the user is verified by the Authority. These authorities then use the certificate to authenticate that the user is indeed the user when another computer attempts to communicate securely with that user. It is critically important that the user be correctly associated with the user's key or sensitive information may be passed to an unauthorized user accidentally. It is also critical that the Certificate Authorities be trustworthy and secure. Many measures are taken to ensure the security of Certificate Authorities including concrete bunkers with steel doors, key cards, security cameras, and alarms.

Future Directions

One of the greatest concerns to cryptographers is the continued increase in computational power of modern computes. Current cryptography relies on the use of very large numbers to make attempts to calculate the hashing algorithm of a public key encryption system impossible. However, advances in quantum physics have led to the creation of quantum computer prototypes. Quantum computers have the potential to be exponentially faster than current computers. They would have the computational power to break nearly any encryption by brute force analysis, simply computing every possible combination and trying it to see if it works. However, quantum computers will likely also have the ability to improve cryptographic methods as well. On balance, computer security may be no better or worse if they are ever produced on a commercial basis.

Summary

Cryptography is the study and practice of writing in secret code or hiding information. Cryptography uses advances in mathematics, computer science, and information theory to improve codes and authentication methods for improved security. Developments in physics, calculus, and information technology are driving an arms race between cryptographers, who encrypt information, and cryptanalysts, who decrypt information to learn the secrets of others. Quantum computing, in particular, will change the balance, at least initially, in favor of the cryptanalysts.

Concept Reinforcement:

1. What are some of the sciences used to advance cryptography?

2. What are the computational and mathematical concepts upon which the privacy of the public encryption key protocol depends?

3. Why are advances in quantum physics of such concern to cryptographers?

Chapter 43 – Telematics

Chapter Objective:

- Understand the significance that science has in telematics, and analyze its impact on future technologies in this field

Introduction

Telematics involves combining telecommunications technology with computers to transmit data over long distances. The term is most commonly used to refer to the combination of global positioning satellites (GPS) with on board computers in automobiles in the form of in-car navigation devices and automated roadside assistance programs like GM's OnStar. The use of communications devices to control computers such as voice command responsive car stereo systems is another example of telematics. Telematics relies on advances in physics, mathematics, electronics, computer science, ergonomics, cognitive ergonomics, and social and behavioral sciences to achieve its ends.

Telematics in the Automobile

Telematics is becoming a common feature of trucks and cars, with GPS navigation systems either preinstalled on high-end luxury vehicles, or as a peripheral device easily mounted on the vehicle by the consumer. GPS navigators communicate with GPS satellites to compute their location and to compute destination arrival time. Miniaturization of electronics devices using advances in physics, materials science, and engineering electronics has allowed the development of these low power, highly graphic, highly dependable devices at low cost.

Human interaction scientists study the design of the telematics devices to ensure ease of use, especially given that many users interact with the devices while driving. Not only must the devices be easy to use physically, but mentally as well. Driver distraction is undesirable.

However, GPS navigation systems and other telematics devices can be an added distraction while driving. Not only are they often placed on the windshield, thereby blocking the driver's vision, they can play music, function as cell phones via Bluetooth technology, and they are highly graphic. Behavioral scientists examine the way in which drivers interact with navigation systems to ensure that their attention remains on the road and not on the navigator.

Because these devices can allow a vehicle to be located at any time, there are several social issues with which social scientists and ethicists must contend. For example, if a car is stolen, it would be possible to locate the car and send police. It is possible to remotely disable the vehicle so that is stops on command. While this is clearly beneficial to the owner, the police, and as a matter of public safety, the potential for government abuse of the technology is of concern. Individuals could be tracked and monitored, located and even harassed if their location were available to malicious users.

Future Developments

As navigation devices become more sophisticated and communications devices become smaller and faster, there is strong potential to integrate these existing technologies with other sensors on vehicles to make them safer. Cars might be able to sense when they are too close to a hazard and warn the driver. Some vehicles are already programmed to take automatic actions, such as slowing down when approaching another vehicle from behind too quickly. The final outcome will be cars capable of driving themselves.

Other uses of telematics include child locator devices. Every year, nearly 750,000 children are reported missing. If children had telematics locator devices, they could call for assistance even if they did not know where they were, and they could be found quickly and easily by law enforcement. Gang-related crime could be reduced if juvenile offenders were outfitted with telematic devices that would communicate not only their whereabouts, but also the identity of others in the area with whom they should not consort.

Summary

Telematics is an interdisciplinary field combining the talents of physicists, mathematicians, materials scientists, behavioral scientists, and computer scientists. Telematics involves combining telecommunications technology with computers to transmit data over long distances. The term is most commonly used to refer to the combination of GPS satellite technology with on board computers in automobiles. But telematics stands poised to expand to cover many more applications of the telecommunications and computer technology disciplines.

Concept Reinforcement:

1. What are some of the sciences used in telematics?

2. What are some of the problems scientists are addressing in telematics?

3. What new applications of telematics may develop as the technology advances and telematic devices are miniaturized further?

Chapter 44 – Leaders in Information Technology

Chapter Objective:

- Understand the significance that science has played in helping the leaders in information technology achieve success

Introduction

Leaders in information technology (IT) possess at least two strengths. One, they solve problems. Two, they have the knowledge in information technology, or are able to gain the knowledge quickly to solve the problems with which they are presented. They must have, or be able to acquire on the job, a strong science background. Many of today's IT leaders have advanced degrees in computer science and engineering. Others, however, have backgrounds in business and learned about IT as part of their job.

Leadership

What is a leader? A leader is a person who solves problems, influences others, sets a good example, is goal oriented, communicates well, and does what is necessary to get the job done. Leaders are not born, they are made. Leadership practices can be learned in formal school settings, informally by observing others, and most importantly by applying what has been learned in daily work situations. Leaders usually learn more from their failures than their successes. To become a leader, become a problem solver. To become a problem solver, a developing leader must have a solid grasp of the key knowledge and concepts underlying the industry. In the IT industry, those topics are computer science, mathematics, logic, physics, electronics, telecommunications, software engineering, and information management.

Educational and Employment Background

To gain perspective of the kinds of knowledge held in common by most IT executives, some examples may be useful.

Dr. Greg Papadopoulos

Sun Microsystems is a major software, systems, and microelectronics company headquartered in Santa Clara, CA. Dr. Greg Papadopoulos serves as the Chief Technology Officer and Executive Vice President of Research and Development. His educational background includes a PhD from MIT in electrical engineering and computer science. He conducted research in computer science as an associate professor at MIT, was a development engineer at Hewlett-Packard and Honeywell, and chief scientist at Boeing and Sun Microsystems. One of his many important contributions to computer design was spearheading the team that designed the CM6 massively parallel supercomputer. It is easy to see that a strong background in computer science, electronics, mathematics, and electrical engineering has helped Dr. Papadopoulos achieve success as a leader in IT.

Dr. Sophie Vandebroek

Born and educated in Belgium, Dr. Sophie Vandebroek received her Ph.D. in electrical engineering from Cornell University in 1990. She is chief technology officer and president of Xerox Innovation Group for Xerox International headquartered in Norwalk, CT. Dr. Vandebroek grew into the position by serving with Xerox for many years in positions of increasing responsibility. She was chief engineer of Xerox Corporation and vice president of the Xerox Engineering Center. She holds 12 US patents, and is a Fellow of the Institute of Electronic and Electrical Engineers, and a Fulbright Fellow. Dr. Vandebroek has capitalized on her education and scientific knowledge to achieve a position of leadership at a major IT company, a position rarely held by a woman.

Rebecca Blalock

Rebecca Blalock is corporate Chief Information Officer and senior vice president at Southern Co. in Atlanta, GA. Unlike the other leaders profiled here, Ms. Blalock's educational background is in business and finance. She had been serving the assistant to the CEO of Southern Company for 6 months when he made her CIO, an area in which she had no experience. She was promoted to the position because of her ability to work with information. She spent a great deal of her time initially in that appointment learning the technical aspects of the position from the personnel in her division. Her advice is to first learn the technology trends, second learn the needs of the business and how IT can address them, and third, build your public speaking skills.

Bill Gates

No discussion of leaders in IT would be complete without including Microsoft's founder, Bill Gates. Mr. Gates' interest developed in high school when he took computer programming classes. He programmed the school's scheduling software and did some programming for several companies until they discovered his age and stopped contracting with him. He attended Harvard from 1973 to 1975, where he majored in computer science. He took a leave of absence from Harvard to focus on his company, Microsoft, and never went back. Although his formal computer science training came to an end, Mr. Gates never stopped learning. In the early years of Microsoft's growth, Mr. Gates was heavily involved in programming and researching the needs of computer users. He studied trends in the industry and remained current on new advances in computer science and systems. His understanding of programming was legendary within Microsoft's programming department and was clearly a factor in his success.

Summary

Regardless of their educational backgrounds, leaders in information technology possess at least two strengths, they solve problems and have the knowledge in information technology, or are able to gain the knowledge quickly to solve the problems with which they are presented. Many of today's IT leaders have advanced degrees in computer science and engineering. Others, however, have backgrounds in business and learned about IT as part of their job. They must have a strong science background or be able to acquire it on the job. Without knowledge of the strengths and limitations of today's hardware and software, they would not be able to plan for and create the next generation of products and applications.

Concept Reinforcement:

1. What are two things all of the leaders in information technology have in common?

2. What are the important characteristics of a leader?

3. What are the topics of which a leader in IT must be knowledgeable?

Chapter 45 – Job Opportunities in Information Technology

Chapter Objective:

- Understand the significance of science in the information technology field and job opportunities

Introduction

According to the US Department of Labor Statistics, the information technology (IT) field is growing at above the average rate of job growth in the US. Employment opportunities for computer software engineers, computer support specialists, customer support representatives, network and computer systems administrators, and telecommunications equipment installers and repairers are available in publishing industries, motion picture and sound recording industries, broadcasting, telecommunications, data processing and hosting, and related services. The vast majority of IT professionals work in private industry.

Education, Training, and Compensation

Almost every position in the IT industry requires at least some education beyond high school. Many of the computer repair technicians, computer operators, telecommunications installers and repairers positions require a vocational certificate that can be earned within a year after high school or on the job training. Little science knowledge is needed to perform the functions of these jobs. However, these are the lowest paid workers in the IT professions. They also have the highest levels of unemployment in the IT professions. The median income (half of the employees make more, half of the employees make less, this is not the same as an average wage) for workers in these categories is approximately $35,000.

Computer software engineers, systems analysts, network and computer systems administrators, and other computer specialists typically possess a bachelors degree or higher. Most of them have degrees in computer science or engineering, although some come from other science backgrounds. Typically their training includes mathematics including calculus, physics, electronics, computer programming, information science, and electrical engineering. They have a high level of science knowledge and the ability to apply that knowledge to problems in the IT profession. They have very low unemployment rates and job growth in their professions is strong. They are also well paid, with median incomes ranging from $65,000 to $88,000 annually. Some of these professionals advance to the level of IT management, where their median salaries exceed $100,000.

Computer scientists and researchers are the most highly educated and trained IT professionals, with most of them holding a Ph.D. in computer science or an engineering field related to computers. Their unemployment levels are low and their median income is over $93,000 per year. These highly trained professionals are educated in advanced computer science, electrical engineering and electronics, physics, materials science, mathematics, and information science. These are the professionals who make the scientific discoveries that in turn generate the next developments in computer technology. Without a strong foundation in science and a true love of science, they would not be able to perform their job functions.

Summary

Careers in IT almost always require at least some education beyond high school. As a result, IT professions of nearly all types offer above average pay scales compared to other professions in the US. As the level of education and science training needed to function in an IT career grows, so does the salary and job security. IT professionals must be prepared not only to obtain an initial education in their chosen career, but to remain up to date as the science behind computer technology advances. They may do so by continuing education either through their own informal learning or through formal coursework.

Concept Reinforcement:

1. What science knowledge is needed for careers in IT?

2. What is the relationship between the level of science knowledge needed and salary scales?

3. What is the relationship between the level of science knowledge needed and unemployment levels?